Black H●les

An Introduction

Second Edition

T0269025

Black Holes

Derek Raine & Edwin Thomas

May 2009

Black Holes

An Introduction

Second Edition

Derek Raine & Edwin Thomas

University of Leicester, UK

Imperial College Press

Published by

Imperial College Press
57 Shelton Street
Covent Garden
London WC2H 9HE

Distributed by

World Scientific Publishing Co. Pte. Ltd.
5 Toh Tuck Link, Singapore 596224
USA office: 27 Warren Street, Suite 401-402, Hackensack, NJ 07601
UK office: 57 Shelton Street, Covent Garden, London WC2H 9HE

Library of Congress Cataloging-in-Publication Data
Raine, Derek J., 1946–
    Black holes : an introduction / Derek Raine & Edwin Thomas. -- 2nd ed.
    p. cm.
    Includes bibliographical references and index.
    ISBN-13 978-1-84816-382-9
    ISBN-10 1-84816-382-7
    ISBN-13 978-1-84816-383-6 (pbk)
    ISBN-10 1-84816-383-5 (pbk)
    1. Black holes (Astronomy)   I. Thomas, E. G. (Edwin George), 1937–   II. Thomas, Edwin.
    III. Title.
    QB843.B55R35 2009
    523.8'875--dc22

                                                            2009022739

British Library Cataloguing-in-Publication Data
A catalogue record for this book is available from the British Library.

Printed in Singapore

Preface

New science can seem quite weird at first: Newton's mystical action-at-a-distance; Maxwell's immaterial oscillations in a vacuum; the dice-playing god of quantum mechanics. In due course however we come to accept how the world is and teach it to our students. Our acceptance, and theirs, comes principally through mastery of the hard details of calculations, not from generalised philosophical debate (although there is a place for that later).

Black holes certainly seem weird. We know they (almost certainly) exist, although no-one will ever 'see' one. And they appear to play an increasingly central role both in astrophysics and in our understanding of fundamental physics. It is now almost 90 years since the Schwarzschild solution was discovered, 80 years since the first investigations of the Schwarzschild horizon, 40 years on from the first singularity theorems and perhaps time that we can begin to dispel the weirdness and pass on something of what we understand of the details to our undergraduate students. This is what we attempt in this book. We have tried to focus on the aspects of black holes that we think are generally accessible to physics undergraduates who may not (or may) intend to study the subject further. Many of the calculations in this book can be done more simply using more sophisticated tools, but we wanted to avoid the investment of effort from those for whom these tools would be of no further use. Those who do go further will appreciate all the more the power of sophistication.

The presentation assumes a first acquaintance with general relativity, although we give a brief recapitulation of (some of) the main points in chapter 1. The treatment is however very incomplete: we do not consider the Einstein field equations because we do not demonstrate here that the black hole geometries are solutions of the equations. In fact, very little prior knowledge of relativity is required to study the properties of given black hole spacetimes. Chapter 2 is devoted to classical spherically symmetric, or non-rotating (Schwarzschild) black holes in the vacuum and chapter 3 to axially symmetric, or rotating (Kerr) ones. It is unfortunate that even simple calculations for Kerr black holes rapidly become algebraically complex. We have tried not to let this obscure the intriguing physics. We ask the reader to stick with it. After all, there may be a lot of it, but it is only elementary algebra. We give only a brief overview of charged black holes. These have played an important role as algebraically simpler models for many of the properties of rotating holes, and they are important as such in higher dimensions, but it is intrinsically difficult to maintain an interest in the physics of objects that probably do not exist, especially since we are going to treat the rotating holes in detail anyway.

In chapter 4 we attempt to explain the quantum properties of black holes without recourse to quantum field theory proper. The calculations here are less rigorous than in the rest of the book (probably a gross understatement) but many variations on the standard theme are now available in textbooks, reviews and lecture notes on the internet and we see no merit in repeating these. We hope our approach is more useful than a crash course in quantum field theory. It does, of course, assume a more than superficial understanding of standard quantum theory. Chapter 6 closes the book with a brief review of black hole astrophysics in so far as it is relevant to the observation of black holes. For this second edition we have added chapter 5 on wormhole metrics and time travel and a set of solutions to the problems. The new edition has also given us the opportunity to revise and clarify some of the text and problems and to add some new problems.

We are aware that we have omitted many contemporary topics in black hole physics, not least the properties of general black holes, perturbation of black holes and the role of black holes in string theories. We regard these as beyond the scope of the book (and in the last case of the expertise of the authors). We hope (and believe) that working out long-hand the details of what we do include will provide a firm foundation for those students who will go on to study such advanced topics and a firm understanding and appreciation of the properties of black holes for those who do not.

Derek Raine
Ted Thomas

Contents

Chapter 1

RELATIVISTIC GRAVITY

1.1 What is a black hole?

Black holes arise because gravity affects the way light waves travel through space. Newtonian dynamics does not treat the effect of gravity on light, but we can use our Newtonian intuition to guess what sort of interaction there might be, given that there is one. The propagation of light is governed by its speed c, so it is natural to look for a characteristic speed in Newtonian gravity. The obvious one is the escape velocity $v_{esc} = (2GM/R)^{1/2}$ from a body of mass M and radius R. So we expect that light emitted from the body will fail to escape to a distant observer when $v_{esc} > c$, that is when the mass of a body of density ρ exceeds the value

$$M \sim (c^2/G)^{3/2}\rho^{-1/2},$$

where we have used the approximation $\rho \sim M/R^3$.

Before the discovery in the twentieth century of white dwarf stars, and later neutron stars, the densest matter known was something like lead. So the mass M_* of the smallest object having a normal density, of around say $\rho_* \sim 5000$ kg m^{-3}, that could prevent its light from escaping to a distant observer would appear to be

$$M_* \sim 10^8(\rho_*/\rho)^{1/2}M_{\odot}, \tag{1.1}$$

where $M_{\odot}$ is the solar mass. The mass M_* is very much greater than that of any known star. Such an object would be invisible or, as we would now say, would be a black hole.

These Newtonian arguments were known to Mitchell (the successor to Cavendish at Cambridge, and the first person to carry out Cavendish's experiment to measure G) and to Laplace, who was the first to carry out detailed studies of the many-body problem in the solar system.

Two things have changed this picture. The first was the discovery of a relativistic theory of gravity, Einstein's general theory of relativity, in which the behaviour of light under the influence of gravity is treated unambiguously. The second was the discovery of matter having densities in the range 10^9 to 10^{17} kg m^{-3} in the form of white dwarfs and neutron stars. The existence of bodies of such high densities suggests that stellar mass black holes might exist. We can add to these another class of astronomical object, the active galactic nuclei (or AGNs), in which the central objects have normal densities but masses of the order of 10^8 $M_\odot$. This combination of mass and density make AGNs candidates for black holes.

In a relativistic theory of gravity, by definition, the local speed of light is always c, the speed of light in a vacuum in the absence of gravity. The Newtonian picture of the emitted light being slowed down and turned back by gravity is therefore not appropriate. However, we can get a truer picture from an application of the equivalence principle. The presence of a gravitational field introduces a relative acceleration between freely-falling frames of reference. The equivalence principle then leads to an approximate relation between time interval $d\tau$ measured at radius R from a body of mass M and the corresponding time interval $d\tau'$ measured at infinity (Will, 1993)

$$d\tau \approx \left(1 - \frac{2GM}{Rc^2}\right)^{1/2} d\tau'. \tag{1.2}$$

(The exact relation depends on the full theory of gravity.) This says that a clock at radius r runs slow compared to a clock at infinity. There is a corresponding redshift z of light emitted at frequency ω and received at ω' given by $\omega/\omega' = 1 + z = d\tau'/d\tau$. This suggests that $z \to \infty$ as the Newtonian potential $2GM/R \to c^2$ and hence that the light from the surface of a body at this potential would be redshifted to invisibility. Thus the body would be a black hole. The relativistic condition $2GM/R = c^2$ is, of course, analytically the same as the Newtonian condition $v_{esc} = c$. But the condition for the Newtonian approximation to be valid is that $GM/R \ll c^2$, so we require the full theory of gravity to investigate this behaviour consistently and to treat black holes correctly. Furthermore, the gravitational potential on the surface of a neutron star is about $0.1c^2$ so we need a general relativistic theory of stellar evolution to be confident of understanding evolution beyond this stage.

The replacement of Newton's theory of gravity by Einstein's general theory of relativity does not alter the relationship between mass and density in equation (1.1), except that now the density is to be interpreted as the average density within the boundary of the (non-rotating) black hole. But it does alter our picture of the spacetime of a black hole and how it gravitates. A relativistic black hole has no material surface; all of its matter has collapsed into a singularity that is surrounded by a spherical boundary called its event horizon. The event horizon is a one-way surface: particles and light rays can enter the black hole from outside but nothing can escape from within the horizon of the hole into the external universe. An outgoing photon that originates outside the event horizon can propagate to infinity but in so doing it suffers a gravitational redshift: in Newtonian language it looses energy in doing work against the gravitational potential. This redshift is larger the closer the point of emission is to the horizon. On the other hand a photon or particle emitted inside the horizon in any direction must inevitably encounter the singularity and be annihilated. (This is strictly true only in the simplest type of black hole: in more general black holes destruction is not inevitable and the fate of a particle or photon inside the hole is more complicated.) A photon emitted at the horizon towards a distant observer stays there indefinitely. For this reason one can think of the horizon as made up of outwardly directed photons.

1.2 Why study black holes?

The importance of black holes for gravitational physics is clear: their existence is a test of our understanding of strong gravitational fields, beyond the point of small corrections to Newtonian physics, and a test of our understanding of astrophysics, particularly of stellar evolution. Current theories show that black holes are an almost inevitable consequence of the way that massive stars evolve: we therefore expect to find black holes amongst the stars in the Galaxy, and it appears that we do.

There is also a surprising and quite unexpected reason why black holes turn out to be important: this is for the potential insight they offer into the connection between quantum physics and gravity. We shall see that black holes appear formally to satisfy the laws of thermodynamics, with Mc^2 in the role of internal energy, the acceleration due to gravity in the guise of temperature and the black hole area as entropy. But this turns out to be more than a formal analogy. When we include the effects of quantum physics we find that black holes behave as real objects with a non-zero temperature and entropy: in particular they radiate like black bodies. The analogy with thermodynamics is therefore not just a formal one, but black holes really do obey the laws of thermodynamics.

We can now turn this argument around. Since black hole radiation involves a mixture of gravity and quantum physics this connection necessarily leads us into the territory of quantum gravity, and, since quantum gravity is the missing link in a complete picture of the fundamental forces, to 'theories-of-everything'. Any theory-of-everything has to be consistent with thermodynamics, and hence with black hole thermodynamics. Therefore any theory-of-everything should be able to predict the thermodynamic properties of black holes from *ab initio* statistical calculations. It is therefore interesting that theories that treat 'strings' as fundamental entities have been partially successful in this regard. It appears that black holes will play a central role in our understanding of fundamental physics.

1.3 Elements of general relativity

It is assumed that the reader has had a first acquaintance with a course on general relativity, for example from one of the many excellent introductory textbooks (for example, Kenyon, 1990, Hartle, 2003). In this section we shall present some of the main ideas of the theory, but only in the form of a brief review.

1.3.1 The principle of equivalence

The principle of equivalence tells us that local experiments (those carried out in the immediate vicinity of an event) cannot distinguish between an accelerated frame of reference and the presence of a gravitational field. In both cases we observe that bodies subject to no non-gravitational forces fall with equal acceleration. (We use the double negative in 'no non-gravitational forces' to emphasise that gravity may or may not be present, but no other forces are acting.) This creates difficulties for the Newtonian approach to dynamics because that requires us to choose a non-accelerated (or 'inertial') frame of reference. In Newtonian physics we get round this problem by designating the distant stars as a non-accelerated reference frame. This is a non-local, non-causal solution and therefore unsatisfactory. It is obviously non-local, and it is non-causal because there is no mechanism by which this reference frame is singled out, except by the fact that it gives the right answers (for example, for the motion of the planets in the Solar System).

Einstein was struck by the observation that all bodies fall with the same acceleration in a gravitational field. Newtonian gravity offers no explanation for this *universality of free-fall.* So Einstein used the universality of free fall to enunciate the principle of equivalence and made this the basis of his general relativistic theory of gravity. The principle of equivalence implies that in free fall we cannot detect the presence of a gravitational field by local experiments. (In free fall all bodies move inertially whether or not gravity is present.) Therefore in a local freely falling frame of reference we already know the laws of physics in the presence of gravity: they are the same as if gravity were absent!

1.3.2   The Newtonian affine connection

From a practical point of view it is not very easy to use the principle of equivalence directly for calculations. This is because in the presence of gravity the local freely falling frame of reference is changing from event to event and we are not experienced at doing calculations in ever changing frames of reference. Rather we need to translate this point of view into a fixed, but arbitrary, reference frame. Although reference frames and coordinate systems are not the same thing, (because the axes of a reference frame are not required to be tangents to coordinate lines), for the present purposes we shall ignore the distinction between them.

Imagine therefore that the system of coordinates $(\xi^0, \xi^1, \xi^2, \xi^3) = (\xi^\mu)$, $(\mu = 0, 1, 2, 3)$ corresponds momentarily to the natural choice of the freely-falling observer, with ξ^0 the Newtonian time (up to a factor of c). Suppose further that another set of coordinates (x^μ) are defined in a global patch (although not necessarily the whole) of spacetime. Each set of coordinates is given in terms of the other by $x^\mu = x^\mu(\xi^\nu)$ and $\xi^\mu = \xi^\mu(x^\nu)$, $(\mu, \nu = 0, 1, 2, 3)$. Note that on the left of these equations the variable stands for an independent coordinate and on the right for a function. The eliding of these distinct meanings by use of the same symbol is common practice and useful for keeping track of dependencies provided that care is taken.

According to Newtonian physics a test body subject to no non-gravitational forces will move along a spacetime trajectory defined by

$$\frac{d^2\xi^\mu}{d\tau^2} = 0. \tag{1.3}$$

Transforming to our (x^μ) coordinates this becomes, after some calculation of partial derivatives,

$$\frac{d^2x^\mu}{d\tau^2} + \frac{\partial x^\mu}{\partial \xi^\alpha}\frac{\partial^2\xi^\alpha}{\partial x^\nu \partial x^\rho}\frac{dx^\nu}{d\tau}\frac{dx^\rho}{d\tau} = 0, \tag{1.4}$$

where we are employing the usual summation convention implying a sum over repeated indices. The Greek indices are assumed to range over the values 0,1,2,3 throughout. To derive (1.4) from Eq. (1.3) we have used

$$\frac{\partial x^\alpha}{\partial \xi^\mu}\frac{\partial \xi^\mu}{\partial x^\beta} = \delta^\alpha_\beta, \tag{1.5}$$

which follows from differentiation of $x^\alpha(\xi^\mu(x^\beta)) = x^\alpha$. Eq. (1.4) is of the form

$$\frac{d^2x^\mu}{d\tau^2} + \Gamma^\mu_{\nu\rho}\frac{dx^\nu}{d\tau}\frac{dx^\rho}{d\tau} = 0. \tag{1.6}$$

This equation for the motion of a test body holds whether or not gravity is present. If gravity is present the only place it can appear in this equation is through the quantity $\Gamma^\mu_{\nu\rho}$, called the affine connection. Looked at in this way, gravity therefore does not enter through an additional force term on the right hand side of (1.3). The only difference between the presence or absence of a gravitational field is that in the latter case it will be possible to recover the Eq. (1.3) everywhere in a single inertial coordinate system, not just locally in a freely-falling frame (because in the absence of gravity a local freely-falling frame is automatically a global inertial frame).

1.3.3 Newtonian gravity

To make the connection with the usual form of the equation of motion of a test body in Newtonian gravity we must be able to choose coodinates in which the affine connection takes an appropriate form. In fact, we must have, in some suitably chosen frame of reference (x^μ),

$$\Gamma^0_{ij} = 0; \quad \Gamma^i_{0j} = 0; \quad \Gamma^i_{00} = \frac{\partial \phi}{\partial x^i},$$

where $i,j = 1,2,3$, and ϕ is the Newtonian gravitational potential, since with these values for $\Gamma^\mu_{\nu\rho}$ we recover from (1.5) the equations of motion in a Newtonian gravitational field:

$$x^0 = ct = c\tau;$$

$$\frac{d^2x^i}{d\tau^2} = -\frac{\partial \phi}{\partial x^i}.$$

This also tells us that the affine connection is related to the distribution of matter through the extension to a general coordinate system of Poisson's equation $\nabla^2\phi = 4\pi G\rho$. Since this involves second derivatives of ϕ the relation between the affine connection and matter must involve the derivatives of the affine connection. The appropriate combinations of derivatives can be shown to be related to the curvature of (Newtonian) spacetime.

Newtonian gravitation is therefore a theory of the structure of spacetime, the relevant structures being the affine connection, the privileged time coordinate t and the Euclidean spatial metric. In Newtonian physics the (affine) geometry of spacetime is measured by the paths of particles and is unrelated to the geometry of time and space as measured by clocks and rods. Relativity is a lot simpler: there is only one geometry. The geometry of time and space, as measured by clocks and rods, itself governs the motion of particles, as we shall explain below.

1.3.4   Metrics in relativity

In a freely-falling frame special relativity is valid *locally* (whether or not gravity is present), and the spacetime interval (the proper distance or proper time) between neighbouring events, ds, is given by the familiar line element (or metric)

$$ds^2 = c^2 dt^2 - (dx^2 + dy^2 + dz^2), \qquad (1.7)$$

where t, x, y, z are the time and rectangular spatial coordinates of the freely falling (inertial) observer. In tensor notation this line element can be written

$$ds^2 = \eta_{\alpha\beta} d\xi^\alpha d\xi^\beta, \qquad (1.8)$$

where $\eta_{\alpha\beta}$ is the metric tensor and we are using $(\xi^\alpha) = (ct, x, y, z)$ for this special coordinate system at a point. With these coordinates the metric has components $\eta_{\alpha\beta} = \mathrm{diag}(1, -1, -1, -1)$. In this convention ds^2 is positive for a timelike interval, zero for a lightlike interval and negative for a spacelike interval. This is convenient for dealing with the motion of particles. For positive ds^2 then $ds/c = d\tau$, where $d\tau$ is the proper time between the events. For negative ds^2 then $(-ds^2)^{1/2} = dl$ is the proper distance between the two events. An alternative convention often used for the metric is $\eta_{\alpha\beta} = \mathrm{diag}(-1, 1, 1, 1)$. Care is needed in copying formulae from references to adjust these, if necessary, to the convention being employed.

In a general coordinate system, which we shall call (x^μ), we have

$$ds^2 = \eta_{\alpha\beta} \frac{\partial \xi^\alpha}{\partial x^\mu} \frac{\partial \xi^\beta}{\partial x^\nu} dx^\mu dx^\nu = g_{\mu\nu} dx^\mu dx^\nu \qquad (1.9)$$

(from the chain rule for partial derivatives, $d\xi^\alpha = (\partial\xi^\alpha/\partial x^\mu)dx^\mu$). The quantities $g_{\mu\nu}$ are called the metric coefficients. They are symmetric, that is to say $g_{\mu\nu} = g_{\nu\mu}$, and are, in general, functions of the spacetime coordinates.

In the absence of gravity we can find a *global* coordinate system (ξ^α) in which the metric takes the form (1.8) everywhere. In the presence of gravity we can find such coordinates only in an infinitesimal neighbourhood of each spacetime point.

Problem 1 *Show that neither the Minkowski metric Eq. (1.7) nor the sum of squares is an invariant under the Galilean transformation and hence that there is no interval in Newtonian spacetime (i.e. Newtonian spacetime does not admit a metric).*

The motion of particles is again governed by Eq. (1.6) in relativity, because the arguments leading to it are still valid, except that now both $t \neq \tau$ and $x^0 \neq c\tau$ (since these would not be compatible with the metric Eq. (1.8) or (1.9)).

Now we see that in local freely falling coordinates the metric is obtained from first derivatives of the ξ^μ whereas the affine connection depends on second derivatives. We therefore expect that there will exist a relation between the components of the affine connection and derivatives of components of the metric. This is indeed the case:

$$\Gamma^\mu_{\nu\rho} = \frac{1}{2} g^{\mu\alpha} \left(\partial_\nu g_{\rho\alpha} + \partial_\rho g_{\nu\alpha} - \partial_\alpha g_{\nu\rho} \right), \tag{1.10}$$

where $\partial_\alpha f = \partial f / \partial x^\alpha$ and $(g^{\mu\nu})$ is the inverse matrix to $(g_{\mu\nu})$, so

$$g^{\mu\alpha} g_{\alpha\nu} = \delta^\mu_\nu.$$

Thus we find the values of $\Gamma^\mu_{\nu\rho}$ from the values of the metric coefficients appropriate to a particular gravitational field.

So in relativity there is just one spacetime geometry defined by the metric and measured by both clocks and rods *and* by particle paths. The motion of particles, is governed by Eq. (1.5), called the geodesic equation (because it is of the same form as the equation for the shortest paths on a curved surface, which are called geodesics). Note that the physical interpretation of the coordinates (x^μ) (what they measure physically) depends on the metric coefficients, so cannot be determined unless the metric is known.

1.3.5 *The velocity and momentum 4-vector*

In this section we define two important 4-vectors relating to the motion of a particle, namely its velocity and momentum; acceleration will be dealt with later. The 4-velocity vector of a particle with position $(x^\alpha(\tau))$ is given by

$$(u^\mu) = \left(\frac{dx^\mu}{d\tau} \right) = \left(\frac{dx^0}{d\tau}, \frac{dx^1}{d\tau}, \frac{dx^2}{d\tau}, \frac{dx^3}{d\tau} \right),$$

formally as in special relativity. The 4-momentum of a particle of mass m_0 is $p^\mu = m_0 u^\mu$.

We lower indices using the metric tensor $g_{\mu\nu}$, so the covariant components of momentum are $p_\mu = g_{\mu\nu}p^\nu$. Conversely, we raise indices using $g^{\mu\nu}$, so the contravariant components are $p^\mu = g^{\mu\nu}p_\nu$.

Taking the scalar product of the 4-velocity with itself gives a relation that we shall use repeatedly below:

$$u^\mu u_\mu = c^2, \tag{1.11}$$

and, similarly,

$$p^\mu p_\mu = m_0^2 c^2. \tag{1.12}$$

Recall that a scalar produce is independent of the frame of reference in which it is evaluated. In particular, we can choose the local freely falling frame of the particle. In this frame special relativity is valid, so the metric takes the form (1.7) and the components of (u^μ) are $(c, 0, 0, 0)$. Thus in this frame we readily verify Eq. (1.11) and similarly Eq. (1.12).

1.3.6   General vectors and tensors

By definition, if we make a coordinate transformation $x \to x'(x)$, then the contravariant components of a 4-vector v^μ transform according to

$$v'^\mu = \frac{\partial x'^\mu}{\partial x^\nu} v^\nu$$

and the covariant components as

$$v'_\mu = \frac{\partial x^\nu}{\partial x'^\mu} v_\nu,$$

with

$$\frac{\partial x'^\mu}{\partial x^\nu} \frac{\partial x^\nu}{\partial x'^\rho} = \delta^\mu_\rho.$$

Higher rank tensors (having multiple indices) transform by extension of this rule to each index. For example:

$$g'_{\mu\nu} = \frac{\partial x^\alpha}{\partial x'^\mu} \frac{\partial x^\beta}{\partial x'^\nu} g_{\alpha\beta}.$$

Problem 2 *Obtain the inverse transformations*

$$v^\mu = \frac{\partial x^\mu}{\partial x'^\nu} v'^\nu \quad and \quad v_\mu = \frac{\partial x'^\nu}{\partial x^\mu} v'_\nu.$$

Problem 3 *By showing that $v_\mu v^\mu = v'_\mu v'^\mu$ verify the invariance of the scalar product.*

1.3.7 Locally measured physical quantities

We often need to calculate the energy or velocity of a particle as measured by a local observer, for example, an observer having a fixed location in some global coordinate system, or a locally freely falling observer. The energy of a particle with respect to a local observer with 4-velocity u^μ_{obs} is the time component of the 4-momentum of the particle in the observer's frame of reference and is therefore obtained by projecting the 4-momentum on to the velocity 4-vector of the observer. So

$$\mathcal{E} = u^\mu_{\text{obs}} p_\mu = m_0 \gamma c^2,$$

where $\gamma = (1 - v^2/c^2)^{-1/2}$ is the usual relativistic gamma factor and v the local velocity of the particle. As in section 1.3.5, we have evaluated the scalar product in the local freely falling frame of the observer where special relativity applies. The scalar product is an invariant, that is, its value is independent of the coordinate system used to evaluate it, so, for example, we can evaluate it in a local inertial frame and also in a global coordinate system and thus relate the global energy to the locally measured energy. We will make frequent use of this property of the scalar product in subsequent chapters.

Physical quantities in two local frames at the same event will be connected by a Lorentz transformation between them even though one or both of the frames may be accelerating. This follows because the instantaneous rates of clocks and lengths of rods are not affected by accelerations and depend only on the relative velocities.

1.3.8 Derivatives in relativity

Given a vector field $v^\mu(x)$, say, it is easy to see that the usual partial derivative $\partial v^\mu / \partial x^\nu$ is, in general, not a physically meaningful quantity: we can give it any value we like by the choice of coordinates x^μ. In particular, a vector field which is constant in one coordinate system, in which its partial derivatives will vanish, may have non-zero derivatives in another coordinate system. The problem arises because the usual definition of a partial derivative compares the values of the quantity in question at two different points and hence, in general, at points where the coordinates can be changed independently. Rates of change having a physical interpretation will be formed from differences in quantities at the same point. Various such derivatives exist, depending on how the quantities are brought to the same point.

The most common, although not the simplest, is the (so called) covariant derivative. This is a generalisation of the behaviour of a tangent vector to a curve. Let $V^\mu = dx^\mu / d\tau$ be the tangent vector to the curve $x^\mu = x^\mu(\tau)$. Eq. (1.6) tells us that

$$\delta V^\mu = -\Gamma^\mu_{\nu\rho} V^\nu \delta x^\rho$$

is the change in the tangent vector on going from x^μ to $x'^\mu = x^\mu + \delta x^\mu$ along a geodesic. We now use this to define the parallel transport of any vector, A^μ say, between two neighbouring points. To form the covariant derivative we take the difference between the vector A^μ at x' and the parallely transported $A^\mu + \delta A^\mu = A^\mu - \Gamma^\mu_{\nu\rho}A^\nu \delta x^\rho$ at x', in the usual limit:

$$\nabla_\nu A^\mu = \lim_{\delta x \to 0} \frac{A^\mu(x+\delta x) - (A^\mu(x) + \delta A^\mu)}{\delta x^\nu}$$
$$= \frac{\partial A^\mu}{\partial x^\nu} + \Gamma^\mu_{\nu\rho}A^\rho.$$

The quantity $\nabla_\nu A^\mu$ is the covariant derivative of A^μ, also denoted by a semi-colon as $A^\mu_{;\nu}$, or sometimes by DA^μ/dx^ν. The covariant derivative therefore measures the derivative corresponding to parallel transport. Note that in this notation the condition for a geodesic becomes that the tangent vector should have zero covariant derivative along the tangent, so $V^\nu \nabla_\nu V^\mu = 0$. This is the total or directional derivative, also written in terms of the affine parameter τ as $DV^\mu/d\tau = 0$.

Problem 4 *The covariant derivative of a scalar is just the usual (partial) derivative (because the value of a scalar is independent of the coordinate system). Hence, for two vector fields, A^μ and B_μ, we have $\nabla_\nu(A^\mu B_\mu) = \partial_\nu(A^\mu B_\mu)$. Use this to show that*

$$\nabla_\nu B_\mu = \frac{\partial B_\mu}{\partial x^\nu} - \Gamma^\rho_{\mu\nu}B_\rho.$$

Since we have seen that in a freely falling frame of reference the components of the affine connection vanish at a point, another definition of the covariant derivative is that it is the ordinary (partial) derivative in a freely falling frame.

A second type of derivative that is sometimes useful is the Lie derivative. To define this we use for the transported vector field at $x'^\mu = x^\mu + V^\mu \delta\tau$ the vector obtained by using the standard transformation law between coordinate systems, namely,

$$A^\mu + \delta A^\mu = \frac{\partial x'^\mu}{\partial x^\nu} A^\nu$$
$$= (\delta^\mu_\nu + \frac{\partial V^\mu}{\partial x^\nu}\delta\tau)A^\nu.$$

Note that the transformation is here regarded as an active one, taking the point x^μ to a different point x'^μ, rather than a passive change of coordinates at a given point. The Lie derivative is then defined as

$$\mathcal{L}_V A^\mu = \lim_{\delta\tau \to 0} \frac{A^\mu(x+\delta x) - (A^\mu(x) + \delta A^\mu)}{\delta\tau}$$
$$= V^\nu \partial_\nu A^\mu - A^\nu \partial_\nu V^\mu.$$

Problem 5 *Show that the partial derivatives in the definition of the Lie derivative can be replaced by covariant derivatives, i.e. that*

$$\mathcal{L}_V A^\mu = V^\nu \nabla_\nu A^\mu - A^\nu \nabla_\nu V^\mu.$$

Roughly speaking we can say that the Lie derivative is the directional derivative of a vector field along a curve adjusted for the change in the tangent.

Problem 6 *Generalise the tensor transformation law, the covariant derivative and the Lie derivative to a tensor field $T^{\mu\nu}$.*

1.3.9 Acceleration 4-vector

As an example of the covariant derivative we can construct the acceleration 4-vector, which is obtained from the velocity 4-vector $(u^{\mu}) = (dx^{\mu}/d\tau)$ as follows:

$$
\begin{aligned}
a^{\mu} &= \frac{Du^{\mu}}{d\tau} = \frac{dx^{\alpha}}{d\tau}\frac{Du^{\mu}}{dx^{\alpha}}\\
&= \frac{dx^{\alpha}}{d\tau}\left(\frac{\partial u^{\mu}}{\partial x^{\alpha}} + \Gamma^{\mu}_{\beta\alpha}u^{\beta}\right)\\
&= \frac{du^{\mu}}{d\tau} + \Gamma^{\mu}_{\beta\alpha}u^{\beta}u^{\alpha}.
\end{aligned}
$$

The magnitude of the rest frame or proper acceleration a of a particle is given by

$$a^{\mu}a_{\mu} = -a^{2}. \tag{1.13}$$

We shall give an example of the use of the Lie derivative below (problem 8).

1.3.10 Paths of light

In relativity the path of a light ray is also governed by the metric through

$$0 = g_{\mu\nu}dx^{\mu}dx^{\nu}$$

and

$$\frac{d^{2}x^{\mu}}{d\lambda^{2}} + \Gamma^{\mu}_{\beta\gamma}\frac{dx^{\beta}}{d\lambda}\frac{dx^{\gamma}}{d\lambda} = 0,$$

in which λ is a parameter which varies along the world line of a light ray (and is called an affine parameter, because it maintains for light rays the usual form of the geodesic equation).

1.3.11 Einstein's field equations

The problem of relating the metric coefficients $g_{\mu\nu}(x^{\alpha})$, as functions of the coordinates, to the distribution of matter is solved by Einstein's field equations , which are

$$R_{\mu\nu} - \frac{1}{2}g_{\mu\nu}R = \frac{8\pi G}{c^{4}}T_{\mu\nu}. \tag{1.14}$$

Here $R_{\mu\nu}$ is the Ricci curvature tensor given by

$$R_{\mu\nu} = \frac{\partial \Gamma^\gamma_{\mu\nu}}{\partial x^\gamma} - \frac{\partial \Gamma^\gamma_{\mu\gamma}}{\partial x^\nu} + \Gamma^\gamma_{\mu\nu}\Gamma^\delta_{\gamma\delta} - \Gamma^\delta_{\nu\gamma}\Gamma^\gamma_{\mu\delta} \tag{1.15}$$

and $R = g^{\alpha\beta}R_{\alpha\beta}$ is the Ricci scalar. The tensor $T_{\mu\nu}$ is the energy momentum tensor which includes all the sources of the gravitational field excluding the energy and momentum in the gravitational field itself which is accounted for by the non-linearity of the equations. Equations (1.14) provide ten non-linear partial differential equations for the metric coefficients. In chapters 2 and 3 we shall be studying certain symmetrical solutions of the Einstein field equations in a vacuum, for which $T_{\mu\nu} = 0$. In chapter 5 we shall need the full equations (1.14).

Problem 7 *Show that if $T_{\mu\nu} = 0$ then $R = 0$.*

1.3.12 Symmetry and Killing's equation

Despite the complexity of Einstein's equations a surprising number of exact solutions are known. These are obtained by imposing symmetries on the spacetime which restrict the possible form of the metric. We shall meet two cases in the following, namely spherical and axial symmetry. In order to exploit these symmetries we shall need one mathematical result which we derive here.

Let the vector k^μ be a direction of symmetry; for example k^μ might point along the azimuthal (ϕ -) direction in spherical symmetry, i.e. $(k^\mu) = (0, 0, 0, 1)$ in spherical polar coordinates. Then we shall show that

$$k_{\mu;\nu} + k_{\nu;\mu} = 0, \tag{1.16}$$

where the semi-colon denotes the covariant derivative

$$k_{\mu;\nu} = \frac{\partial k_\mu}{\partial x^\nu} - \Gamma^\lambda_{\mu\nu}k_\lambda.$$

Eq. (1.16) is called Killing's equation.

Let events P at (x^μ) and Q at (x'^μ) be separated by a small distance in the direction of symmetry k^μ, so

$$x'^\mu = x^\mu + \varepsilon k^\mu. \tag{1.17}$$

In moving from P to Q we do two things: we change the label of the point from x^μ to $x^\mu + \varepsilon k^\mu$ and we make an active transformation of coordinates from x^μ to x'^μ. The metric coefficients at x'^μ are related to those at x^μ by

$$g_{\mu\nu}(x^\lambda) = \frac{\partial x'^\alpha}{\partial x^\mu}\frac{\partial x'^\beta}{\partial x^\nu}g'_{\alpha\beta}(x'^\lambda).$$

As we have moved in a direction of symmetry the *function* $g'_{\mu\nu}(x'^\lambda)$ of the new coordinates has the same form as the *function* $g_{\mu\nu}(x^\lambda)$ of the original coordinates, so we can substitute

$$g'_{\mu\nu}(x'^\lambda) = g_{\mu\nu}(x'^\lambda).$$

Expanding $g_{\alpha\beta}(x'^\lambda)$ as a Taylor series and using Eq. (1.17) to evaluate the partial derivatives we get

$$g_{\mu\nu}(x^\lambda) = (\delta^\alpha_\mu + \varepsilon\partial_\mu k^\alpha)(\delta^\beta_\nu + \varepsilon\partial_\nu k^\beta)(g_{\alpha\beta}(x^\lambda) + \varepsilon k^\lambda \partial_\lambda g_{\alpha\beta}(x^\lambda) + ...).$$

Multiplying out the brackets to first order in ε gives us

$$g_{\mu\nu}(x^\lambda) = g_{\mu\nu}(x^\lambda) + \varepsilon k^\lambda \partial_\lambda g_{\mu\nu}(x^\lambda) + \varepsilon g_{\mu\beta}(x^\lambda)\partial_\nu k^\beta + \varepsilon g_{\alpha\nu}(x^\lambda)\partial_\mu k^\alpha.$$

But

$$\partial_\nu(g_{\mu\beta}k^\beta) = \partial_\nu k_\mu = g_{\mu\beta}\partial_\nu k^\beta + k^\beta \partial_\nu g_{\mu\beta},$$

and we get

$$0 = \varepsilon(\partial_\mu k_\nu - k^\alpha \partial_\mu g_{\alpha\nu}) + \varepsilon(\partial_\nu k_\mu - k^\beta \partial_\nu g_{\mu\beta}) + \varepsilon k^\lambda \partial_\lambda g_{\mu\nu}$$
$$= \varepsilon(k_{\mu;\nu} + k_{\nu;\mu}),$$

where the final equality is obtained using the relation Eq. (1.10). A vector field k^μ satisfying this relation is called a Killing vector.

Problem 8 *Show that Killing's equation (1.16) is equivalent to $\pounds_k g_{\mu\nu} = 0$. (You will need the result of problem 6.)*

Chapter 2

SPHERICAL BLACK HOLES

In this chapter we look at the vacuum gravitational field produced by a spherical (non-rotating) mass. The spacetime metric for this case, obtained by Karl Schwarzschild in 1916, was the first exact solution of Einstein's equations to be found, although its properties were not fully understood until much later.

We investigate the geometry in the vicinity of the mass M by studying the motion of test particles and light rays. If the space is a vacuum down to the Schwarzschild radius $r = 2GM/c^2$ we obtain the metric of a spherical black hole. In the next chapter we shall study the metric of a rotating (non-spherical) black hole. The surface $r = 2GM/c^2$ constitutes the event horizon of the (spherical) black hole and acts as a one-way membrane for particles and light. We explore the motion of particles and light rays as a spherical mass collapses to form a black hole and what an observer falling into a spherical black hole would experience. To explore the geometry in the vicinity of the event horizon and the maximum extension of the geometry from the external region we introduce various alternative coordinate systems.

2.1 The Schwarzschild metric

The general spherically symmetric metric can be written in the form

$$ds^2 = g_{00}(t,r)c^2dt^2 + g_{11}(t,r)dr^2 - r^2(d\theta^2 + \sin^2\theta d\phi^2). \tag{2.1}$$

We adopt the timelike convention for the signs of the metric coefficients $(+,-,-,-)$, so $ds^2 > 0$ is a timelike interval and $ds^2 < 0$ is a spacelike interval (section 1.3.4). The metric contains two unknown functions of r and t, g_{00} and g_{11}, to be determined from Einstein's field equations.

In the vacuum of space surrounding a massive spherical body Einstein's equations force the functions to be independent of time, and to take the form

$$g_{00} = \left(1 - \frac{2GM}{c^2r}\right), \tag{2.2}$$

$$g_{11} = -\left(1 - \frac{2GM}{c^2r}\right)^{-1}, \tag{2.3}$$

where M is a parameter and G the gravitational constant. Thus the metric becomes

$$ds^2 = \left(1 - \frac{2GM}{rc^2}\right)c^2dt^2 - \left(1 - \frac{2GM}{rc^2}\right)^{-1} dr^2 - r^2(d\theta^2 + \sin^2\theta d\phi^2). \qquad (2.4)$$

This is the Schwarzschild metric in Schwarzschild coordinates (t, r, θ, ϕ). It describes the space-time in the exterior vacuum region only. Inside a material body the metric is different and depends on the properties of the body such as the equation of state of the material.

The contravariant components of the metric $g^{\mu\nu}$ satisfy

$$g^{\mu\nu}g_{\nu\rho} = \delta^\mu_\rho,$$

defining $(g^{\mu\nu})$ as the matrix inverse to $(g_{\nu\rho})$. Since the metric is diagonal the matrix inverse is readily found, giving

$$g^{00} = \left(1 - \frac{2GM}{c^2 r}\right)^{-1},$$

$$g^{11} = -\left(1 - \frac{2GM}{c^2 r}\right),$$

$$g^{22} = -\frac{1}{r^2},$$

$$g^{33} = -\frac{1}{r^2 \sin^2\theta}.$$

In relativity the interpretation of the coordinates is obtained from the properties of the metric. We give the interpretation of the coordinates and of the parameter M in the next section.

2.1.1 Coordinates

The notation has been chosen here to be suggestive. The coordinate r does indeed play the role of a radial measure; in fact, the sphere $r = $ constant, $t = $ constant has the (spacelike) metric

$$dl^2 = r^2(d\theta^2 + \sin^2\theta d\phi^2), \qquad (2.5)$$

from which it is clear that the element of area on the surface is $rd\theta \times r\sin\theta d\phi$, and hence that the area of the surface is $4\pi r^2$. Thus the r coordinate is related to the proper area, that is the area as measured physically by, say, the diminution in intensity of a wave front of light. The angles θ and ϕ are the usual spherical polar angles. Nevertheless, in relativity one should always be aware that a suggestive notation can sometimes be misleading. Later on we shall see that there are regions of space-time (those within $r = 2GM/c^2$) where r is not a spacelike coordinate and t is not timelike. We shall also find that it is sometimes convenient to use other coordinates to describe the Schwarzschild space-time geometry.

2.1.2 Proper distance

The radial increment of proper distance is obtained from the metric by setting $dt = d\theta = d\phi = 0$; it is not dr, but $dr/(1 - 2GM/rc^2)^{1/2}$. The coordinate r therefore does *not* measure proper distance, as would be determined, for example, by a ruler.

2.1.3 Proper time

Similarly, the time coordinate t is *not* the time kept by a standard clock in the exterior gravitational field of the body. We can see this from the metric (2.4). An increment of proper time kept by a stationary clock at a radius r, for which $dr = d\theta = d\phi = 0$, is related to the increment of coordinate time by

$$d\tau = \left(1 - \frac{2GM}{rc^2}\right)^{1/2} dt. \tag{2.6}$$

The increments $d\tau$ and dt increasingly agree as $r \to +\infty$. Thus the time coordinate t has the interpretation that it is the proper time kept by a clock at an infinite distance from the body, or a close approximation to that time kept by clocks at large distances.

For the purpose of visualisation we can imagine that the body is surrounded by static shells at each value of the radial coordinate r. A standard clock fixed to a shell measures proper time on that shell, and, of course, standard clocks are identical in that they run at the same rate when they are at the same radius. In addition, we imagine a second clock on each shell that has been adjusted to run at the same rate as the distant clock. This second group of clocks measure coordinate time and run faster than the standard clocks by a factor $(1 - 2GM/rc^2)^{-1/2}$. This slowing down of local clocks relative to a distant clock is the effect of gravitational time dilation. The closer we approach $r = 2GM/c^2$, the faster the coordinate clock runs relative to local proper time, until at $r = 2GM/c^2$ it is running infinitely faster.

The Schwarzschild coordinates (t, r, θ, ϕ) provide a global reference frame that enables us to evaluate quantities in terms of the distance and time measurements of an observer at infinity. As we shall see, the values obtained by a distant observer for time intervals, for energies and for velocities of particles differ from local measurements of these quantities.

2.1.4 Redshift

A consequence of this difference in the measurement of time locally and at infinity is that radiation sent out from a radius r is redshifted when received by a static observer further out. Since the wavelength of radiation is proportional to the period of vibration, equation (2.6) gives us

$$\lambda = \left(1 - \frac{2GM}{rc^2}\right)^{1/2} \lambda_\infty,$$

for the relation between the wavelength λ of radiation emitted at r and the wavelength λ_∞ received at infinity. For the redshift z, defined by

$$z = \frac{\lambda_\infty - \lambda}{\lambda},$$

we get

$$1 + z = \left(1 - \frac{2GM}{rc^2}\right)^{-1/2}.$$

2.1.5 Interpretation of M and geometric units

The quantity M in the Schwarzschild metric arises as a constant of integration in the solution of the Einstein equations. It is in fact the mass of the central gravitating body. This is determined by a comparison of the motion of a body in the Schwarzschild metric with the motion of a body in the Newtonian gravitational field of a mass M, which we shall carry out below. For convenience we define a *geometric mass* m,

$$m = \frac{GM}{c^2}. \tag{2.7}$$

The geometric mass has the dimension of a length, so in SI units, M would be given in kilograms and m in metres.

To simplify further, we replace ct by t and $c\tau$ by τ. The new time is therefore also measured in length units, hence, if we use SI units, in metres. Thus, we have $t(\text{in metres}) = c \times t(\text{in seconds})$, and similarly for τ, but instead of this cumbersome notation we allow the context to distinguish which t and τ we mean. In geometrical units the metric is

$$d\tau^2 = \left(1 - \frac{2m}{r}\right) dt^2 - \left(1 - \frac{2m}{r}\right)^{-1} dr^2 - r^2(d\theta^2 + \sin^2\theta d\phi^2). \tag{2.8}$$

2.1.6 The Schwarzschild radius

The radial coordinate value $r = 2GM/c^2 = 2m = r_S$ is called the Schwarzschild radius. We see that something strange appears to happen to the Schwarzschild metric at this value of r, since the metric coefficient g_{11} becomes infinite. For a normal gravitating body the Schwarzschild radius lies within the object where the vacuum Schwarzschild solution ceases to apply. For the Sun, for example, we get $r_S \sim 3$ km. So, for the dynamics of the Solar System, or even for motion around a neutron star, the problem can be safely ignored. But the question of what happens at the Schwarzschild radius leads to the subject of black holes. Thus, if a spherical mass has radius $R > r_S$, then the Schwarzschild metric in the region $r > R$ gives its exterior gravitational field; if the region $r \geq r_S$ is a vacuum then the Schwarzschild metric for $r > r_S$ gives the exterior gravitational field of a black hole.

Inside $r = r_S$ the Schwarzschild metric is again a solution of Einstein's vacuum field equations and gives the interior geometry of a black hole. As we shall show later, the reason that we cannot follow what happens at the Schwarzschild radius $r = r_S$ in Schwarzschild coordinates is because the coordinates are invalid at this surface, not because of any physical limitation.

At $r = 0$ the coefficient of g_{00} becomes infinite. In contrast to the apparent singularity at the Schwarzschild radius, this singularity cannot be removed by a different choice of coordinates. It is a physical singularity where spacetime curvature becomes infinite and timelike worldlines terminate.

2.1.7 *The event horizon*

As we shall see, of the many interesting properties of the surface $r = r_S$ the most novel is that it acts as a one-way membrane for information: things can fall into the region $r < r_S$ but cannot emerge from it. For this reason it is called an event horizon and the region of space within it is called a black hole.

Problem 9 *Show that the (proper) circumference of a black hole of mass m is $4\pi m$ and that the (proper) area is $16\pi m^2$.*

2.1.8 *Birkoff's theorem*

The coefficients of the Schwarzschild metric do not depend on t; such a metric is said to be *stationary*. Furthermore, there are no cross term involving spacelike and timelike increments (such as $dt d\phi$) in the metric. A stationary metric satisfying this further condition is said to be *static*. We have seen that the Einstein equations, which led from the general form (2.1) to the particular form (2.4) or (2.8), imply that a spherically symmetric vacuum solution is static. This result is known as Birkoff's Theorem. It means that if the Sun were to oscillate while remaining spherical the surrounding vacuum would still be described by the Schwarzschild metric (2.4) and therefore that the oscillation could not be detected through the gravitational field. This accords with Newtonian gravity where a spherical distribution of matter gives the same gravitational field as a point of the same mass at its centre. One can emphasize the status of the theorem by rephrasing it as a uniqueness result: the Schwarzschild geometry is the unique spherically symmetric vacuum gravitational field satisfying Einstein's equations.

2.1.9 *Israel's theorem*

A converse to Birkoff's theorem can also be shown to hold with some additional technical assumptions: a static vacuum gravitational field must be spherical in general relativity, hence must be describable by the metric (2.4). We shall explore this in more detail later. Stated as a uniqueness theorem this becomes: the Schwarzschild geometry is the unique static vacuum gravitational field in general relativity. One is always free to choose different coordinates in relativity, so this result does not say that any spherically symmetric vacuum solution must look like (2.8)! But any such solution can be put into the form (2.8) by a suitable transformation of coordinates.

2.2   Orbits in Newtonian gravity

The simplest way to derive the equations of a test body orbiting in the gravitational field of a spherical mass M in Newtonian theory is to use the conservation of energy and angular momentum. (The energy and angular momentum are constants of the motion.) Since the orbit must lie in a plane (by conservation of angular momentum, or, more directly, by symmetry) we choose the equatorial plane, $\theta = \pi/2$, of a spherical polar coordinate system for convenience.

2.2.1   Newtonian Energy

For a particle of unit mass in orbit in the equatorial plane the conservation of energy gives

$$E_{\mathrm{N}} = \frac{1}{2}\left(\left(\frac{dr}{dt}\right)^2 + r^2\left(\frac{d\phi}{dt}\right)^2\right) - \frac{GM}{r},\tag{2.9}$$

where $E_{\mathrm{N}} = $ constant is the total energy per unit mass comprising the kinetic energy of radial and circular motion and the gravitational potential energy.

2.2.2   Angular momentum

Conservation of angular momentum implies (i) that the orbit lies in a plane and (ii) that the magnitude of the angular momentum per unit mass

$$L_{\mathrm{N}} = r^2\frac{d\phi}{dt}\tag{2.10}$$

is constant.

2.2.3   The Newtonian effective potential

Putting together (2.9) and (2.10) we get

$$\frac{1}{2}\left(\frac{dr}{dt}\right)^2 = E_{\mathrm{N}} - \frac{L_{\mathrm{N}}^2}{2r^2} + \frac{GM}{r} = E_{\mathrm{N}} - V_{\mathrm{N}},\tag{2.11}$$

where V_{N} is called the effective potential. (It is *not* the usual gravitational potential itself because it contains the additional centrifugal term. In fact, it is the potential in a frame rotating with the orbit.)

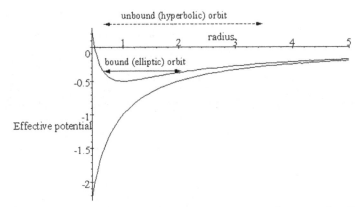

Figure 1 Examples of the Newtonian Effective Potential as a function of radius for zero angular momentum (lower curve) and non-zero angular momentum (upper curve).

2.2.4 Classification of Newtonian orbits

Figure 1 shows a graph of the Newtonian effective potential as a function of radius r in the two cases $L_N = 0$ and $L_N = 1$ for a fixed central mass m. In interpreting the figure remember a horizontal line in the figure represents the motion of a particle (having constant total energy) and that the difference between the total energy and a given curve is the kinetic energy as a function of radius for given angular momentum: there is no implication that the motion is radial!

- For $L_N \neq 0$ we see that $\dot{r}^2$ is positive if $E_N > V_{\min}$. Thus the radial velocity is non-zero in these regions.

- From the figure we can see that orbits with $E_N > 0$ extend to arbitrarily large values of r. These are therefore the unbounded (hyperbolic) orbits.

- Orbits with $E_N < 0$ (but $> V_{\min}$) exist between fixed values of r; these are the bound (elliptic) orbits.

- There is one special case when $E_N = V_{\min}$ where the value of r is fixed at $r_{\text{circ}} = L_N^2/GM$, which is clearly a circular orbit.

Note that, for a central point mass, if $L_N > 0$ a particle with any positive value of energy will be deflected by the potential back to infinity.

2.3 Particle orbits in the Schwarzschild metric

We are now ready to consider the paths of bodies which are subject to no non-gravitational forces, i.e. the trajectories in spacetime followed by bodies in free fall.

These are given by geodesics of the metric (2.4). There are several ways to obtain such particle paths. The simplest is to make use of the symmetries to derive constants of the motion.

2.3.1   Constants of the motion

Let k^μ be a vector in a direction of symmetry, for example $k^\mu = (1,0,0,0)$ along the t axis. Let $u^\mu = dx^\mu/d\tau$ be the tangent vector to the particle path $x^\mu = x^\mu(\tau)$ in spacetime and consider the component of k^μ along the path, namely $u_\mu k^\mu$. The rate of change of $u_\mu k^\mu$ in the direction u^μ is

$$u^\nu(k^\mu u_\mu)_{;\nu} = u^\nu u_{\mu;\nu}k^\mu + u^\nu u^\mu k_{\mu;\nu}$$
$$= u^\nu u_{\mu;\nu}k^\mu + \frac{1}{2}u^\nu u^\mu(k_{\mu;\nu} + k_{\nu;\mu}).$$

Suppose now that the path is a geodesic. Then $u^\nu u_{\mu;\nu} = 0$ because this is the geodesic equation. Furthermore, it can be shown that $k_{\mu;\nu} + k_{\nu;\mu} = 0$ for a vector along a symmetry direction (see chapter 1). Thus $u^\nu(k^\mu u_\mu)_{;\nu} = 0$ and therefore $k^\mu u_\mu$ is constant along a geodesic (because, in words, the equation says that the derivative of $k^\mu u_\mu$ in the direction of the tangent vanishes).

2.3.2   Conserved Energy

Consider first the time direction, $k^\mu = (1,0,0,0)$. This is fairly obviously a symmetry of the static metric, since the metric coefficients are independent of t. (More formally, the metric is unchanged by the transformation $x^\mu \to x^\mu + \varepsilon k^\mu$.) We therefore have

$$g_{\mu\nu}k^\mu u^\nu = g_{00}k^0 u^0 = g_{00}u^0 = E = \text{constant.} \qquad (2.12)$$

because the term in $k^0 = 1$ provides the only non-zero contribution to the sum. Note that we write this expression in terms of u^μ not u_μ because we know that $u^\mu = dx^\mu/d\tau$. Note that $u_0 = g_{0\nu}u^\nu = g_{00}u^0$, so (2.12) is equivalent to

$$u_0 = E \qquad (2.13)$$

(or $u_0 = E/c$ in physical units). Hence, from Eq. (2.12), along any geodesic in the Schwarzschild geometry, we have explicitly

$$(1 - 2m/r)\frac{dt}{d\tau} = E. \qquad (2.14)$$

The tangent vector u^μ to a geodesic is the 4-velocity vector of a particle in free-fall that follows the geodesic. Equivalently, u^μ is the momentum 4-vector of a unit mass particle. To find the meaning of the constant E consider the particle at infinity, then in this limit (2.14) becomes

$$\frac{dt}{d\tau} = E.$$

Now recall from section 1.3.7 that the scalar product $u^{\mu}_{obs}u_{\mu}$ is the energy of a particle having 4-velocity u^{μ} in the frame of a local observer whose 4-velocity is u^{μ}_{obs}. For a stationary observer at infinity $(u^{\mu}_{obs}) = (1, 0, 0, 0,)$ so for a particle at infinity

$$u^{\mu}_{obs}u_{\mu} = \frac{dt}{d\tau} = E.$$

Thus E is the relativistic energy per unit mass of the particle relative to a stationary observer at infinity. The observer at infinity - sometimes called the bookkeeper - is an important notion. This observer uses the global Schwarzschild coordinates (r, θ, ϕ) and coordinate time t, the time kept by clocks at infinity. For this observer the energy of a free particle is conserved.

2.3.3 Angular momentum

Similarly, for the symmetry in the ϕ direction, we have $k^{\mu} = (0, 0, 0, 1)$ from which,

$$g_{\mu\nu}k^{\mu}u^{\nu} = g_{33}k^3u^3 = g_{33}u^3 = -L = \text{ constant.} \tag{2.15}$$

Note that $u_3 = g_{3\nu}u^{\nu} = g_{33}u^3$ so (2.15) is equivalent to

$$u_3 = -L. \tag{2.16}$$

Using $u^3 = d\phi/d\tau$ and the Schwarzschild metric coefficient $g_{33} = -r^2$ this is explicitly

$$r^2\frac{d\phi}{d\tau} = L = \text{ constant.} \tag{2.17}$$

In this case we interpret conservation of the ϕ component of momentum as the conservation of angular momentum, with L the angular momentum per unit mass relative to an observer at infinity. This interpretation is clearly borne out by comparison with the equivalent Newtonian equation. (Replace proper time τ by Newtonian time t and note that $r \times r\dot{\phi}$ is the angular momentum per unit mass.) This explains the choice of minus sign in (2.16).

Since the gravitational field is spherically symmetric, any free-fall orbit must lie in a plane. (Because all directions are equivalent, there is no way to choose a particular direction out of the plane in which to move. This is not true if the body is acted on by forces, such as rocket motors, because these can define a preferred direction.) As in the Newtonian case we take $\theta = \text{constant} = \pi/2$, the equatorial plane, so $\sin\theta = 1$, which turns out to be the most convenient choice.

Note that we prefer to use energy and angular moment *per unit mass* and hence 4-velocity rather than 4-momentum for particles, since particles of different mass follow the same worldline. We use 4-momentum whenever we want to compare the behaviour of a particle as it tends towards the speed of light with that of a photon.

2.3.4   The effective potential

Our final equation to determine the fourth coordinate along a geodesic comes from the Schwarzschild metric itself. Dividing (2.8) by $d\tau^2$ we get

$$1 = \left(1 - \frac{2m}{r}\right)\left(\frac{dt}{d\tau}\right)^2 - \left(1 - \frac{2m}{r}\right)^{-1}\left(\frac{dr}{d\tau}\right)^2 - r^2\left(\frac{d\phi}{d\tau}\right)^2. \qquad (2.18)$$

These equations, (2.14), (2.17) and (2.18), can be rearranged to give

$$\left(\frac{dr}{d\tau}\right)^2 = E^2 - \left(1 + \frac{L^2}{r^2}\right)\left(1 - \frac{2m}{r}\right) = E^2 - V_{\text{eff}}^2(r). \qquad (2.19)$$

The quantity $V_{\text{eff}}(r)$, defined by this equation, is known as the effective potential.

Note that the relation (2.18) (and hence (2.19)) is equivalent to the relation between energy and momentum for a relativistic particle of unit mass, $u_\mu u^\mu = p_\mu p^\mu = 1$, as demonstrated in the following problem.

Problem 10 *Derive equation (2.19) starting from the scalar product of the 4-velocity of the particle $g^{\mu\nu}u_\mu u_\nu = 1$.*

Henceforth we shall denote the mass of a test particle by m_0. This should not be confused with the geometric mass m of a black hole. Bringing together the results of the last sections we see that equations (2.14), (2.17) and (2.19) govern the behaviour of particle orbits in the Schwarschild spacetime. We shall now look at some applications of these equations.

2.3.5   Newtonian approximation to the metric

We are now going to show how the Newtonian equations of motion for a body moving slowly (with respect to light) in an equatorial orbit can be obtained from an approximation to the Schwarzschild metric. Putting $\theta = \pi/2$ and dividing through the Schwarzschild metric by dt^2 we get

$$\left(\frac{d\tau}{dt}\right)^2 = 1 - \frac{2m}{r} - \left(1 - \frac{2m}{r}\right)^{-1}\left(\frac{dr}{dt}\right)^2 - r^2\left(\frac{d\phi}{dt}\right)^2. \qquad (2.20)$$

In the Newtonian limit velocities are small, so we can neglect the factor $(1 - 2m/r)^{-1}$ which multiplies $(dr/dt)^2$. (This factor represents a small correction to an already small quantity.) Note that this is equivalent to ignoring the factor multiplying dr^2 in the metric. We then set $E^2 \approx 1 + 2E_N$, where E_N is small. (In physical units this is $E^2 \approx c^2 + 2E_N$.) We also need to use here the relation (2.14) for $d\tau/dt$ which, to the same order of approximation, is

$$\left(\frac{d\tau}{dt}\right)^2 \approx \frac{\left(1 - \frac{2m}{r}\right)^2}{1 + 2E_N} \approx \left(1 - \frac{4m}{r}\right)(1 - 2E_N) \approx 1 - \frac{4m}{r} - 2E_N,$$

neglecting products of small terms. With these approximations (2.20) gives the Newtonian equation (2.11). Since E_N is the Newtonian energy (per unit mass), E is clearly the relativistic energy (per unit mass) for a slowly moving particle.

We see that the metric

$$d\tau^2 = \left(1 - \frac{2m}{r}\right) dt^2 - dr^2 - r^2(d\theta^2 + \sin^2\theta d\phi^2) \tag{2.21}$$

reproduces the Newtonian equations of motion for slowly moving bodies. We can therefore regard this as the Newtonian approximation to the metric (even though Newtonian gravity is not a metric theory, so does not strictly correspond to a spacetime structure with any metric.) We see that in this approximation the gravitational potential produces a redshift, but not a curvature of space. This metric cannot account correctly for the perihelion precession of Mercury or for the bending of light.

Problem 11 *Verify directly from the geodesic equations (1.6) that the Newtonian metric (2.21) reproduces the Newtonian equations of motion for slowly moving bodies. (You will need to compute the components of the affine connection as in chapter 1 and neglect small quantities.)*

2.3.6 Classification of orbits

Figure 2 shows a plot of the effective potential $V_{\text{eff}}(r)$ against r. There are still regions of bound and unbound orbits, although there is no reason to believe these will be elliptical or hyperbolic now. Between the two there is a marginally bound (but not parabolic) orbit and the special case of a circular bound orbit. The most striking feature however, is that no amount of angular momentum can keep an orbit of sufficient energy out of the region $r < 2m$. Even worse, for particles with small angular momentum relative to the mass of the central object, in fact if $L < 2\sqrt{3}m$, all orbits end up inside $r = 2m$. Compare this with the Newtonian case (Fig. 1) where any angular momentum prevents a particle reaching $r = 0$. This is the first hint that if we can extend the vacuum field to such small radii then peculiar things will happen.

2.3.7 Radial infall

Let a particle be in radial free fall. Then its 4-velocity is $(u^\alpha) = (u^0, u^1, 0, 0)$ for purely radial motion. From (2.14) we have

$$u^0 = \frac{dt}{d\tau} = \frac{E}{\left(1 - \frac{2m}{r}\right)},$$

and, using $1 = g_{\alpha\beta}p^\alpha p^\beta$, we have

$$u^1 = \pm\left(E^2 - 1 + \frac{2m}{r}\right)^{1/2}. \tag{2.22}$$

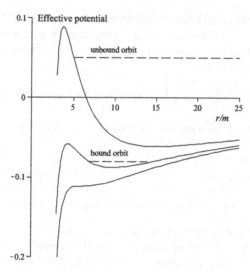

Figure 2 Examples of the relativistic effective potential, plotted as $\frac{1}{2}\left[V_{eff}^2 - 1\right]$ as a function of radius for $L/m = 4.3, 3.75, 3.46$.

Problem 12 *Obtain the result (2.22).*

The simplest orbit we can treat is the radial infall of a particle released from rest at infinity. In this case the particle at infinity has rest energy only, so $E = 1$. Hence $u^1 = \left(\frac{2m}{r}\right)^{1/2}$.

Problem 13 *Show that the escape velocity from radius r measured by a stationary observer at r is $\left(\frac{2m}{r}\right)^{1/2}$. What is the escape velocity from the horizon at $r = 2m$?*

2.3.8 The locally measured energy of a particle

We have just seen that for a distant observer the total energy per unit mass of the infalling particle has a constant value E. For comparison, the energy $\mathcal{E}$ measured by a local observer maintained at fixed coordinate position r with 4-velocity $(u^\mu) = (dt/d\tau, 0, 0, 0)$ is given by $u^\mu p_\mu$, the projection of the particle 4-momentum along u^μ.

For an observer at rest in the Schwarzschild spacetime, from the metric,

$$d\tau^2 = (1 - 2m/r)dt^2,$$

so

$$(u^\mu) = ([1 - 2m/r]^{-1/2}, 0, 0, 0)$$

and

$$\mathcal{E} = u^\mu p_\mu = \frac{m_0 E}{\left(1 - \frac{2m}{r}\right)^{1/2}}. \tag{2.23}$$

It follows that as measured by local stationary observers the energy of the freely falling particle increases with decreasing r.

How is this compatible with the view of distant observers, who assign a fixed value $m_0 E$ to the energy? Equation (2.23) shows that the two energies are related by a redshift factor. It is helpful to understand why this is just what is required by energy conservation. Suppose the infalling particle is stopped by a stationary observer, its energy $\mathcal{E}$ converted to radiation and sent back to infinity. The energy of this radiation certainly suffers a redshift, so the energy received at infinity is $\mathcal{E}\,(1 - 2m/r)^{1/2}$. Unless this is equal to the energy $m_0 E$ that we started with, we should be able to construct a perpetual motion machine. Thus, (2.23) exactly expresses energy conservation.

We can obtain one more result from this argument. The kinetic energy of the particle in the local frames of reference is obtained by subtracting the rest mass energy from the total energy:

$$\mathcal{E} - m_0,$$

where as usual we put $c = 1$. Suppose that the particle is brought to rest at r and just the kinetic energy sent back to infinity as radiation. This energy suffers a redshift relative to the distant observer, so the energy received at infinity is

$$(\mathcal{E} - m_0)\left(1 - \frac{2m}{r}\right)^{1/2} = m_0 E - m_0\left(1 - \frac{2m}{r}\right)^{1/2}.$$

Thus, according to the distant observer, after being brought to rest the particle is left with an energy

$$m_0\left(1 - \frac{2m}{r}\right)^{1/2}, \tag{2.24}$$

this being the difference between the initial energy at infinity $m_0 E$ and the amount received back in radiation. As $r \to 2m$ the energy sent to infinity is just the original total energy $m_0 E$. The particle cannot give up any more energy to infinity than this. For suppose that its rest mass is converted to radiation at the horizon and sent back to infinity. According to (2.24) the energy received at infinity is redshifted to zero.

2.3.9 Circular orbits

Bringing an infalling particle to rest at a given radius, as in the previous section, is somewhat artificial. More realistically we can consider the circular orbits that are possible about the central mass.

For a circular orbit ($r = $ constant) the two conditions

$$\frac{dr}{d\tau} = 0 \quad \text{and} \quad \frac{d^2 r}{d\tau^2} = 0 \tag{2.25}$$

must be satisfied. The first condition, in conjunction with Eq. (2.19), tells us that $E = V(r)$, so the energy lies on the effective potential curve. The radial equation

(2.19) gives also, in general, by differentiation with respect to τ,

$$\frac{d^2r}{d\tau^2} = -\frac{1}{2}\frac{d}{dr}V_{\text{eff}}^2(r).$$

Hence, the second condition of (2.25) gives

$$\frac{dV_{\text{eff}}^2}{dr} = 2V_{\text{eff}}\frac{dV_{\text{eff}}}{dr} = 0,$$

and hence, finally, $dV_{\text{eff}}/dr = 0$. Therefore circular orbits are possible only at turning points of the effective potential.

Using the explicit form for $V_{\text{eff}}^2(r)$ from (2.19) we find the turning points occur at

$$r = \frac{L^2}{2m} \pm \frac{1}{2}\sqrt{\frac{L^4}{m^2} - 12L^2}. \qquad (2.26)$$

For $L^2 > 12m^2$ there are two solutions, the negative sign in (2.26) corresponding to a maximum of $V_{\text{eff}}(r)$, which is an unstable point, and the positive sign to a minimum of $V_{\text{eff}}(r)$, which is therefore stable.

Problem 14 *Show that the solutions obtained by taking the $\pm$ signs in (2.26) correspond respectively to a minimum and a maximum of the effective potential.*

At $L^2 = 12m^2$ there is just one circular orbit: the maximum and minimum of the curve come together at a point of inflection for this value of L. This orbit is marginally stable ($d^2V_{\text{eff}}/dr^2 = 0$) and is the innermost (marginally) stable orbit. From (2.26) the condition $L^2 = 12m^2$ gives a radius $r = 6m$ for this innermost stable orbit. Because the orbit is only marginally stable, a particle at $r = 6m$ perturbed inward by a small amount will fall towards $r = 2m$. (At this stage we cannot say what will happen after that.)

2.3.10   Comparison with Newtonian orbits

In Newtonian gravity circular orbits can exist at any radius, however small, and are always bound. The angular momentum per unit mass of a particle in an orbit at r is $L = (mr)^{1/2}$ and its binding energy per unit mass is $E = -m/2r$. A particle moving in from infinity must lose energy before it can go into orbit, and an infinite amount of energy can be extracted from a particle falling into a 'Newtonian' black hole.

For a Schwarzschild black hole stable circular orbits exist for $r \geq 6m$ only, and they are all bound. The angular momentum per unit mass of a particle in a circular orbit at radius r is

$$L = \left(\frac{mr}{1 - 3m/r}\right)^{1/2}$$

and the energy per unit mass is

$$E = \frac{1 - 2m/r}{(1 - 3m/r)^{1/2}}. \tag{2.27}$$

For $r < 6m$ circular orbits are unstable. In the range $4m < r < 6m$ the unstable orbits are bound ($E < 1$). At $r = 4m$ we have $E = 1$ and $L = 4m$. In the range $3m \leq r \leq 4m$ the orbits are unstable and unbound ($E \geq 1$). Thus, a particle coming in from infinity with $E \geq 1$ can (in principle) settle into an unstable circular orbit without losing any energy, provided that it has the appropriate value of angular momentum. Of course, since the orbit is unstable, in practice a small perturbation would eject the particle again. Nevertheless, this is a non-Newtonian feature of the strong gravitational field.

As $r \to 3m$ both E and L tend to infinity, so only zero mass particles (photons) can orbit at this radius.

Problem 15 *Show that the specific angular momentum of a particle in a circular orbit is given by*

$$L = \left(\frac{mr}{1 - 3m/r} \right)^{1/2}.$$

Show that the energy of this particle is

$$E = \left(1 - \frac{2m}{r} \right) \left(1 - \frac{3m}{r} \right)^{-1/2}.$$

Problem 16 *Show that the proper period of the orbit is*

$$\tau = 2\pi \left(\frac{r^3}{m} \right)^{1/2} \left(1 - \frac{3m}{r} \right)^{1/2}$$

and that the coordinate period is

$$T = 2\pi \left(\frac{r^3}{m} \right)^{1/2}.$$

Problem 17 *Show that there exists an unstable circular orbit into which a particle coming from infinity with specific energy $E = 1$ and impact parameter $L/E = 4m$ is inserted.*

2.3.11 Orbital velocity in the frame of a hovering observer

What is the orbital speed of a particle as measured by a hovering observer? We can calculate this as follows. The hovering observer has velocity 4-vector

$$(u_H^\mu) = \left((1 - 2m/r)^{-1/2}, 0, 0, 0 \right),$$

and the orbiting particle has 4-velocity

$$(u_\mu) = (E, 0, 0, L),$$

where E is, as usual, the conserved energy per unit mass at infinity. Let the energy per unit mass of the particle be γc^2 with respect to the hovering observer. Then, with $c = 1$, the projection of the momentum of the particle on the 4-velocity of the observer gives

$$\gamma = u_\mu u^\mu = \frac{E}{(1 - 2m/r)^{1/2}}.$$

But recall from special relativity that, in terms of the velocity v of the orbiting particle (and with $c = 1$), $\gamma = (1 - v^2)^{-1/2}$. So using (2.27) for E and solving for v we find

$$v = \frac{(m/r)^{1/2}}{(1 - 2m/r)^{1/2}}.$$

Problem 18 *Show that the local orbital velocity in a circular orbit at radius r equals the escape velocity, $v_{esc} = (2m/r)^{1/2}$, when $r = 4m$ and exceeds the escape velocity for $r < 4m$. Hence deduce that a circular orbit at $r < 4m$ is unbound.*

2.3.12 Energy in the last stable orbit

It is of interest to find the energy of a particle in the innermost stable orbit at $r = 6m$, because this will tell us how much energy can be extracted, in principle, from a particle spiralling slowly inwards. For a circular orbit, we have from (2.19)

$$E^2 = \left(1 - \frac{2m}{r}\right)\left(1 + \frac{L^2}{r^2}\right).$$

With $L^2 = 12m^2$ and $r = 6m$ this becomes

$$E^2 = 8/9.$$

Now a particle at rest at infinity has $E = 1$ (again from (2.19)). So the binding energy per unit mass of a particle in the innermost stable orbit is

$$1 - \sqrt{\frac{8}{9}} = 0.0572,$$

or about 5.7 per cent of $m_0 c^2$ for a particle of mass m_0.

This means that matter spiralling inwards in, for example, an accretion disc (chapter 6), will release 5.7 per cent of its rest mass energy before it plunges into the hole from the last stable orbit. For comparison note that nuclear fusion yields about 0.7 per cent of rest mass energy.

2.4 Orbits of light rays

The behaviour of light can also be used to show the effect of the strong gravitational field at small radii. To obtain the trajectories of light rays consider again the equations of motion for a particle of mass m_0. For conservation of energy we have from (2.14)

$$m_0 \left(1 - 2m/r\right) \frac{dt}{d\tau} = m_0 E.$$

For a photon we let $d\tau \to 0$ and $m_0 \to 0$ with $d\lambda = d\tau/m_0$ finite. Then for a ray of light

$$(1 - 2m/r) \frac{dt}{d\lambda} = E_{ph},$$

where $m_0 E \to E_{ph}$ as $m_0 \to 0$ and $E \to \infty$. Similarly, from Eq. (2.17), conservation of angular momentum gives

$$r^2 \frac{d\phi}{d\lambda} = L_{ph}.$$

where $m_0 L \to L_{ph}$. In addition, we now have

$$d\tau^2 = 0,$$

or, equivalently $g^{\mu\nu} p_\mu p_\nu = 0$, from which

$$\left(\frac{dr}{d\lambda}\right)^2 = E_{ph}^2 - \frac{L_{ph}^2}{r^2}\left(1 - \frac{2m}{r}\right) = E_{ph}^2 - V_{ph}^2, \tag{2.28}$$

where λ is an affine parameter along a ray. The corresponding effective potential $V_{ph}(r)$ has a maximum at $r = 3m$, which is independent of L_{ph}. This is therefore a circular orbit for photons, but because V_{ph} is a maximum the orbit is unstable. The potential $V_{ph}(r)$ has no minima, so there are no stable circular orbits for photons. In other words, although in principle the gravity of a spherical mass can act on light to keep it moving in a circle, in practice this cannot happen.

   Drawing together the above results, the equations (2.14), (2.17) and (2.28) govern the behaviour of light rays in the Schwarzschild geometry.

Problem 19 *Evaluate $d\tau/m_0$ in an inertial frame and show that in the limit as $d\tau \to 0$ and $m_0 \to 0$, $d\tau/m_0 \to dt/E_{ph}$.*

Problem 20 *The local speed of light in relativity is always c. Verify that this is true for the Schwarzschild metric. (Hint: obtain an expression for the local velocity of a finite mass particle and consider the limit as its rest mass tends to 0.)*

2.4.1 Radial propagation of light

The 4-momentum of a light ray propagating radially is $(p^\alpha) = (dt/d\lambda, \pm dr/d\lambda, 0, 0)$. Thus, from (2.4)

$$p^0 = \frac{E_{ph}}{\left(1 - \frac{2m}{r}\right)}$$

where E_{ph} is the conserved photon energy in the Schwarzschild frame. Now from $g_{\alpha\beta}p^\alpha p^\beta = 0$, we obtain $p^1 = \pm E_{ph}$. So

$$(p^\alpha) = \left(\frac{E_{ph}}{\left(1 - \frac{2m}{r}\right)}, \pm E_{ph}, 0, 0\right).$$

We have seen that in section 2.1.4 that light of wavelength λ sent from a stationary source at coordinate r to an observer at infinity suffers a redshift

$$\lambda_\infty = \lambda \left(1 - \frac{2m}{r}\right)^{-\frac{1}{2}}.$$

Now we consider the case of a probe in radial free-fall transmitting signals back to a spaceship at infinity. We want to find the relationship between the received and transmitted wavelengths. This can be done by evaluating the scalar product $u^\alpha p_\alpha$, where u^α is the 4-velocity of the probe and p^α is the 4-momentum of a photon. Evaluating the scalar product in a locally freely falling frame in which the probe is stationary gives

$$u^\alpha p_\alpha = h\nu,$$

where ν is the frequency of the signal in the rest frame of the probe. Now we evaluate the scalar product in the Schwarzschild frame, which is the frame of an observer at infinity. The components of the 4-velocity of the probe in the Schwarzschild frame, assuming it has dropped from rest at infinity, are

$$(u^\alpha) = \left(\left(1 - \frac{2m}{r}\right)^{-1}, -\left(\frac{2m}{r}\right)^{\frac{1}{2}}, 0, 0\right),$$

where u^0 is given by equation (2.14) with $E = 1$. The u^1 component is obtained from $g_{\mu\nu}u^\mu u^\nu = 1$ with the assumption that the probe falls from rest at infinity. The covariant components of the outwardly propagating photon are

$$p_\alpha = \left(h\nu_\infty, -\frac{h\nu_\infty}{\left(1 - \frac{2m}{r}\right)}, 0, 0\right).$$

So

$$u^\alpha p_\alpha = \frac{h\nu_\infty}{\left(1 - \frac{2m}{r}\right)} + \left(\frac{2m}{r}\right)^{\frac{1}{2}} \frac{h\nu_\infty}{\left(1 - \frac{2m}{r}\right)}$$

$$= h\nu_\infty \left(1 - \left(\frac{2m}{r}\right)^{\frac{1}{2}}\right)^{-1} = h\nu,$$

where the second equality follows from the invariance of the scalar product. So finally we get for the wavelength at infinity

$$\lambda_\infty = \lambda \left(1 - \left(\frac{2m}{r}\right)^{1/2}\right)^{-1}. \tag{2.29}$$

Problem 21 *A spaceship plunges from rest at infinity into a black hole. Show that the frequency at which signals sent from infinity with frequency ν_∞ are received by the spaceship at the horizon is $\nu_\infty/2$. What is the frequency of this signal seen by the crew of the spaceship from inside the horizon (assuming that they survive for long enough)?*

2.4.2 Capture cross-section for light

The bending of light by gravity means that non-radial light rays can be captured by a black hole. Later we shall discuss the interior spacetime of a black hole $r < 2m$. For this we shall need to know which light rays can enter this region, that is, we shall need the capture cross-section of the black hole region for light rays. To look at the general properties of the orbits of light rays we return to the corresponding effective potential (Fig. 3)

$$V_{\text{eff}}^2 = \frac{L_{ph}^2}{r^2}\left(1 - \frac{2m}{r}\right).$$

The height of the maximum at $r = 3m$ is $L_{ph}^2/27m^2 = V_{\text{max}}^2$. For $E_{ph} > V_{\text{max}}$ an incoming photon enters $r = 2m$ so it is captured by the hole. For $E_{ph} < V_{\text{max}}$ an incoming photon is scattered by the potential back to infinity, so is not captured. The condition of capture is therefore $L_{ph}/E_{ph} < \sqrt{27}m$.

If the impact parameter of the photon that is just captured by the hole is b, then the capture cross-section of the hole is πb^2. So we have to relate the geometrical quantity b to the dynamical properties of the orbit, L_{ph} and E_{ph}. Now, the angular impact parameter b: $L_{ph} = pb$. For a photon we have $E_{ph} = p$ (recall that $c = 1$ in our units), so we can put $b = L_{ph}/E_{ph}$. This gives us the capture cross-section

$$\sigma = \pi b^2 = 27\pi m^2,$$

whereas the geometrical cross-section is $\pi r_s^2 = 4\pi m^2$.

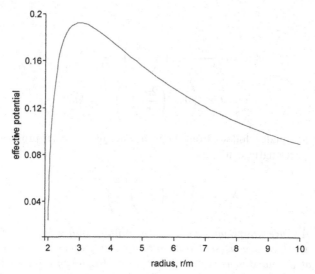

Figure 3   Example of the relativistic effective potential for light rays.

Problem 22 *Using equation (2.28) and the conditions for a circular orbit (section 2.3.9) show that a light ray coming in from infinity with impact parameter $b = 3\sqrt{3}m$ enters an unstable circular orbit of radius $r = 3m$.*

It follows from problem 22 that a light ray coming from the unstable circular orbit at $r = 3m$ and travelling outwards will have an impact parameter $3\sqrt{3}m$. Thus to a distant observer the apparent radius of the black hole is $3\sqrt{3}m$.

Problem 23 *The black hole at the centre of our Galaxy has a mass of $\sim 4 \times 10^6 M\odot$. What is its apparent angular diameter as seen from the Earth, a distance of 8 kpc from the Galactic centre? This is of interest because it is now possible to resolve structures of the order of 40 micro arc seconds using very long baseline interferometry at wavelengths of the order of 1.3mm. (Doeleman et al., 2008)*

2.4.3   The view of the sky for a stationary observer

Suppose that a stationary observer at radius r looking at the sky in the equatorial plane $\theta = \pi/2$ receives a ray of light making an angle ψ' with the outward radial direction. This ray undergoes a proper radial displacement $dr/\left(1 - \frac{2m}{r}\right)^{1/2}$ and a proper tangential displacement $rd\phi$ for each increment $d\lambda$ of affine parameter. From Fig. 4 the angle ψ' is therefore given by

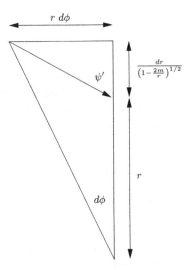

Figure 4   Geometry of a light ray.

$$\sin^2 \psi' = \frac{r^2 d\phi^2}{r^2 d\phi^2 + dr^2 / \left(1 - \frac{2m}{r}\right)}$$

$$= \frac{\left(1 - \frac{2m}{r}\right)}{\left(1 - \frac{2m}{r}\right) + \frac{1}{r^2}\left(\frac{dr}{d\phi}\right)^2}.$$

But $dr/d\phi = (dr/d\lambda)/(d\phi/d\lambda) = (dr/d\lambda)/(L_{ph}/r^2)$, and therefore from (2.28)

$$\left(\frac{dr}{d\phi}\right)^2 = \frac{E_{ph}^2}{L_{ph}^2}r^4 - \left(1 - \frac{2m}{r}\right)r^2,$$

and finally

$$\sin\psi' = \pm\left(1 - \frac{2m}{r}\right)^{1/2}\frac{b}{r}$$

where $b = L_{ph}/E_{ph}$ is the impact parameter. We take the positive square root so that ψ' is near 0 for small b. Then, also,

$$\cos\psi' = \left[1 - \left(1 - \frac{2m}{r}\right)\frac{b^2}{r^2}\right]^{1/2}. \qquad (2.30)$$

Now, according to the calculation of the previous section, an observer close to the horizon at $r = 2m$ will receive only those rays with an impact parameter $b \leq \sqrt{27}m$. The corresponding maximum angle is given by

$$\sin\psi'_{\max} \approx \left(1 - \frac{2m}{r}\right)^{1/2}\frac{\sqrt{27}}{2}$$

as $r \to 2m$. Thus the black hole fills the whole sky except for a disc of angular diameter $\sqrt{27}\left(1 - \frac{2m}{r}\right)^{1/2}$, which closes up to zero for the observer at the horizon.

2.5 Classical tests

We have considered the special cases of radial and circular orbits of particles; now we look at more general orbits. The orbit equations (2.17) and (2.19) allow us to find $dr/d\phi$ as a function of r and hence to compute the paths of bound particles in space. The results can be compared with observation, for example with the perihelion precession of the planet Mercury. Similarly the equations of light rays allow calculation of the bending of light by, for example, the Sun. In these and other ways relativity can be compared with observation. In all of these tests the general relativistic result is found to be at the centre of ever-narrowing error bars. This is discussed further in standard texts on relativity.

These solar system tests involve weak gravitational fields so do not directly imply that strong field effects are correctly predicted by relativity. One example of a strong field test is the speeding up in the period of the binary pulsar (Taylor 1994). But we also look for direct or indirect observation of black holes, having the predicted properties, to verify the theory. (See chapter 5.).

Problem 24 *Obtain the equation of the orbit of a test mass with specific angular momentum L in the equatorial plane of a spherical mass m in the weak field limit $GM/Rc^2 << 1$*

$$\frac{d^2u}{d\phi^2} + u = \frac{m}{L^2} + 3mu^2,$$

where $u = 1/r$. By seeking an approximate solution of the form $u = l^{-1}(1 + \varepsilon \cos \lambda\phi)$ for small eccentricity ε, show that the perihelion rotates by $6\pi m/l$ per orbit.

Obtain similarly the equation of motion of a light ray

$$\frac{d^2u}{d\phi^2} + u = 3mu^2.$$

Show that in the absence of the non-Newtonian term on the right of this equation the motion of light is a straight line. With this term, $3mu^2$, included, verify that there is a unique circular orbit for photons.

2.6 Falling into a black hole

From an astrophysical point of view we are interested in the collapse of a sphere of matter that cannot support itself against gravity. This was first discussed by Oppenheimer and Snyder (1939). From Birkoff's theorem on the uniqueness of the Schwarzschild metric we already know that the collapse of a spherical star from the point of view of an external observer is described by the Schwarzschild geometry.

To be precise let us assume that the pressure support in the star has been removed completely and that the surface of the star is in free fall. This gives us a simple way of discussing the approach of the surface towards $r = 2m$. Specifically, we want to know how much time it takes, according to an observer located at a large fixed value of r, for the surface to approach $r = 2m$.

2.6.1 Free-fall time for a distant observer

For radial free fall we can refer to section 2.3.7 to obtain (2.31) below, or we start again from (2.19),

$$\left(\frac{dr}{d\tau}\right)^2 = E^2 - \left(1 - \frac{2m}{r}\right),$$

where we have put $L = 0$ for radial infall. Assuming that the collapse started with a negligible inward speed,

$$\frac{dr}{d\tau} \to 0 \text{ as } r \to \infty,$$

so $E^2 = 1$. Thus

$$\left(\frac{dr}{d\tau}\right)^2 = \frac{2m}{r}. \tag{2.31}$$

However, viewing the collapse from a large distance, we do not record the proper time of the infalling matter, but we use our own proper time, which is the coordinate time t. Thus we use

$$\frac{dt}{d\tau} = \frac{1}{1 - \frac{2m}{r}},$$

from which

$$\frac{dr}{dt} = -\left(\frac{2m}{r}\right)^{1/2}\left(1 - \frac{2m}{r}\right) \tag{2.32}$$

describes the infall of the surface as seen from a distance. We integrate this from a large value of r, $r = R$, say, to close to $r = 2m$ (using r' as the integration variable to distinguish it from the infall radius):

$$t = -\int_R^r \left(\frac{r'}{2m}\right)^{1/2} \frac{r'dr'}{r' - 2m}.$$

To approximate the integral note that the major contribution is from close to $r' = 2m$ so we put $r' = 2m + \varepsilon$ and hence, expanding in powers of ε,

$$\left(\frac{r'}{2m}\right)^{1/2} \frac{r'}{r' - 2m} = \frac{2m}{\varepsilon} + ...,$$

where the higher order terms not shown are small compared to $1/\varepsilon$. Hence,

$$t \sim -2m \int_R^r \frac{dr'}{r' - 2m},$$

38   Spherical Black Holes

approximately, and
$$t \sim -2m \log \left(\frac{r}{2m} - 1 \right),$$
the contribution from the lower limit $r' = R$, which we have neglected, being much
smaller than the term we retain provided that r is close enough to $2m$. This is the
time taken for the collapsing surface to approach $r = 2m$ from a large radius. Notice
that it tends to infinity as $r \to 2m$.

2.6.2   Light-travel time

But the observer will not see the approach until a photon from the surface has had
time to reach him. Therefore, we have to add the time taken by a ray of light to
reach a large distance from close to $r = 2m$.

For a light ray $d\tau^2 = 0$ so, from the metric with $d\theta = d\phi = 0$ also
$$\frac{dt}{dr} = \frac{1}{1 - \frac{2m}{r}}.$$

Integrating
$$t' = \int_r^R \frac{r'dr'}{r' - 2m} \sim -2m \log \left(\frac{r}{2m} - 1 \right),$$
again taking the main contribution to the integral to come from close to $r = 2m$.

The observer sees the surface at time $T = t + t'$ where, therefore,
$$T = -4m \log \left(\frac{r}{2m} - 1 \right). \qquad (2.33)$$

This tends to infinity at $r = 2m$. Thus, for an external observer the infall to $r = 2m$
takes an infinite time.

2.6.3   What the external observer sees

The wavelength λ_∞ of radiation received at large r coming from a freely falling source
is
$$\lambda_\infty = \lambda \left(1 - \left(\frac{2m}{r} \right)^{\frac{1}{2}} \right)^{-1}.$$
See section 2.4.1 So the red shift is
$$1 + z = \frac{\lambda_\infty}{\lambda} = \left(1 - \left(\frac{2m}{r} \right)^{\frac{1}{2}} \right)^{-1} = \left(\frac{2m}{r} \right)^{-\frac{1}{2}} \left(\left(\frac{r}{2m} \right)^{\frac{1}{2}} - 1 \right)^{-1}.$$

To express this in terms of the time T of a distant observer rewrite equation
(2.33) as follows:
$$\left(\left(\frac{r}{2m} \right)^{\frac{1}{2}} + 1 \right) \left(\left(\frac{r}{2m} \right)^{\frac{1}{2}} - 1 \right) = exp \left(-\frac{T}{4m} \right).$$

Hence

$$1 + z = \left(\frac{2m}{r}\right)^{-\frac{1}{2}} \left(\left(\frac{r}{2m}\right)^{\frac{1}{2}} + 1\right) exp\left(\frac{T}{4m}\right).$$

So as $r \to 2m$ the external observer sees the redshift tend to infinity exponentially quickly as the surface of the star approaches $r = 2m$.

Similarly the luminosity decreases proportionally to $1/(1+z)^2$ (one factor of $1+z$ coming from the redshift and one from time dilation: the photon energy falls and so does the rate of arrival of photons). Thus an external observer sees the luminosity of the star decline in time as $exp(-T/2m)$ as it collapses to $r = 2m$. (A more accurate calculation would take into account non-radial rays leading to a luminosity that falls approximately as $exp(-T/3\sqrt{3}m)$.)

Seen by a distant observer the velocity dr/dt of an infalling object increases during the initial stages of the fall but later, as it approaches the horizon, it decreases to hover forever just outside the horizon. This assumes the object is viewed through infinitely sensitive instruments that can continue to detect it despite its exponentially fading luminosity. Thus, from the point of view of an external observer a collapsing star would never form a black hole. This caused a lot of confusion in the 1960s and led Russian astronomers to call black holes 'frozen stars'. That black holes do form can be seen by looking at the collapse in the reference frame of the falling object.

2.6.4   An infalling observer's time

From equation (2.31) in terms of the proper time of an infalling observer we have

$$\tau = -\int_R^{2m} \left(\frac{r}{2m}\right)^{1/2} dr = \frac{4}{3}m\left(\left(\frac{R}{2m}\right)^{3/2} - 1\right)$$

as the time to fall from $r = R$ to $r = 2m$. Thus the infalling observer reaches $r = 2m$ from a finite distance in a finite proper time.

A falling object can also be viewed by observers situated on static shells that it passes on its way. The velocity of the object relative to the shell at coordinate r is given by

$$v_{loc} = \frac{\text{proper distance}}{\text{proper time}} = \frac{dr}{(1 - 2m/r)\,dt}.$$

Using (2.32) for an object falling from rest, we get

$$v_{loc} = \left(\frac{2m}{r}\right)^{1/2}. \tag{2.34}$$

(An equivalent result was obtained in section 2.3.7.) Thus the locally measured velocity of infall increases steadily with decreasing r and in the limit as $r \to 2m$ we get $v_{loc} \to 1$ (i.e. in physical units the locally measured speed tends to c).

Problem 25 *a) A particle falls from rest at infinity. Show that the time it takes to fall from $r = 2m$ to $r = 0$ is $4m/3$.*
b) Show that the maximum time that a particle can take in going from $r = 2m$ to $r = 0$ is $m\pi$. (Hint: use the metric and the condition that the world line of the falling particle must be light-like in the limit.)
c) Show that it corresponds to a particle falling from rest at $r = 2m$.
d) Why would you expect this world line of maximum proper time to be a geodesic?

2.6.5   What the infalling observer feels

Seen from a local frame fixed with respect to the Schwarzschild coordinates a radially infalling particle having $E = 1$ falls with speed $v_{loc} = (2m/r)^{1/2}$ (section 2.31, or Eq. (2.34)). If we work from the frame of the freely falling particle, the speed of the coordinate lattice in this frame is also $v_{loc} = (2m/r)^{1/2}$. Let τ be the proper time of the particle. The acceleration of the stationary frame is $dv_{loc}/d\tau = (dv_{loc}/dr)(dr/d\tau) = (dv_{loc}/dr)v_{loc} = m/r^2$.

As in Newtonian physics the dependence of the acceleration on radial distance means that an extended body will feel a tidal force. We want to calculate this tidal force from the point of view of the infalling observer (who will after all be subject to the force). We follow the derivation in Taylor and Wheeler (2000). Any two observers agree on their relative acceleration, so the relative acceleration of the head and the feet of a radially aligned infalling observer is

$$dg = \frac{2m}{r^3}dr.$$

To get the tidal force we need the proper distance corresponding to a coordinate distance dr (because our height that we measure in metres is our proper height in free fall, not, in principle, a coordinate displacement in Schwarzschild coordinates, although the two will turn out to be numerically the same).

In a stationary frame the proper distance corresponding to a displacement dr is $dr\,(1 - 2m/r)^{-1/2}$. For the freely falling observer moving with speed $v_{loc} = (2m/r)^{1/2}$, this length is contracted by a factor

$$1/\gamma = \left(1 - v_{loc}^2\right)^{1/2} = (1 - 2m/r)^{1/2}.$$

So the proper length in the freely falling frame is

$$\gamma^{-1}dr\,(1 - 2m/r)^{-1/2} = dr.$$

The tidal acceleration on a body of extension Δr is therefore approximately $(2m/r^3)\Delta r$. At some point this will begin to cause pain for the infalling observer. We want to calculate how long this pain must be endured before the singularity at $r = 0$ is encountered.

Suppose that our criterion for the onset of pain is $\Delta g = g_E$, the gravity at the Earth's surface. This will occur at a radius $r_p = (2m\Delta r/g_E)^{1/3}$. The infall time to $r = 0$ is obtained by integrating (2.31):

$$\tau = -\int_{r_p}^{0} \left(\frac{r}{2m}\right)^{1/2} dr = \frac{2r_p^{3/2}}{3(2m)^{1/2}} = \frac{2}{3}\left(\frac{\Delta r}{g_E}\right)^{1/2}.$$

Note that this is independent of the black hole mass.

Problem 26 *Show that for a standard observer of height 2m the infall time from the threshold of pain to death at the singularity is about 0.3s. Show that for an observer falling into a $1 M_\odot$ black hole the threshold of pain occurs well outside the Schwarzschild radius. For what mass black hole does the pain threshold coincide with the Schwarzschild radius?*

Problem 27 *What mass of black hole would allow an observer in free fall to survive for a year inside the horizon before being destroyed?*

2.7 Capture by a black hole

In section 2.4.2 we looked at the probability that a ray of light would be captured by a black hole taking into account the bending of light by the gravitational field of the black hole. It is of interest to calculate the corresponding capture cross-section of a black hole for particles. The method is similar: any particle with energy $E > V_{\text{eff}}^{\max}$ is captured by the hole. This gives us a critical angular momentum for capture. Instead of angular momentum (for a unit mass particle) L we use the impact parameter b defined in terms of the particle velocity at infinity v by $L = vb$. The capture cross-section is πb^2. With a lot of algebra it can be shown that the maximum of

$$V_{\text{eff}}^2 = \left(1 - \frac{2m}{r}\right)\left(1 + \frac{L^2}{r^2}\right)$$

occurs for $r = (L^2/2m)\left[1 - (1 - 12m^2/L^2)^{1/2}\right]$ and is given by

$$(V_{\text{eff}}^{\max})^2 = \frac{1}{54}\left[l^2 + 36 + (l^2 - 12)\left(1 - \frac{12}{l^2}\right)^{1/2}\right], \tag{2.35}$$

where $l = L/m$.

For the rest of the calculation we consider two limiting cases.

2.7.1 Case I: Capture of high angular momentum particles

Inserting the condition $l \gg 1$ into (2.35) and using the binomial expansion gives

$$(V_{\text{eff}}^{\max})^2 \approx \frac{l^2 + 9}{27}.$$

The critical angular momentum below which capture will occur is therefore $l_*^2 = 27E^2 - 9$. Thus, for a particle of energy E per unit mass and momentum p per unit mass, the critical impact parameter is

$$b_* = \frac{L_*}{p} = \frac{ml_*}{(E^2 - 1)^{1/2}},$$

and the cross-section is therefore

$$\sigma = \pi b_*^2 = \frac{\pi m^2 l_*^2}{(E^2 - 1)}.$$

For large E $(E \gg 1)$ we have $(E^2 - 1)^{-1} = E^{-2}(1 - 1/E^2)^{-1} \approx E^{-2}(1 + 1/E^2)$, so

$$\sigma \approx \frac{\pi m^2}{E^2} \left(1 + \frac{1}{E^2}\right) \left(27E^2 - 9\right) = 27\pi m^2 \left(1 + \frac{2}{3E^2}\right).$$

Note that as $E \to \infty$, $\sigma \to 27\pi m^2$, and we recover the capture cross-section for a photon.

2.7.2   Case II: Capture of low energy particles

For a particle with low energy $(v \ll 1)$ we put $(V_{\text{eff}}^{\max})^2 = E^2 \approx (1 + \frac{1}{2}v^2)^2 \approx 1 + v^2$ as the limit for capture and solve for the critical angular momentum l_* in terms of v. This gives

$$l^2 \approx 16 + 27v^2 + \ldots,$$

neglecting terms in v^4 and higher. Thus

$$\sigma = \frac{16\pi m^2}{v^2},$$

which agrees with our intuition that a particle moving slowly enough will be captured by the gravity of the black hole.

2.8   Surface gravity of a black hole

In this section we are going to derive an important quantity associated with a black hole, called its surface gravity by analogy with the acceleration due to gravity g of Newtonian theory. First we need to determine the acceleration of a particle at rest with respect to a local observer at rest, and then to determine how this acceleration appears from infinity.

2.8.1   The proper acceleration of a hovering observer

The proper or rest frame acceleration a of a particle is related to its 4-acceleration a^μ by equation (1.13):

$$a^\mu a_\mu = -a^2,$$

where

$$a^\mu = \frac{du^\mu}{d\tau} + \Gamma^\mu_{\rho\sigma} u^\rho u^\sigma.$$

The components of the 4-velocity of a hovering observer are

$$u^\mu = \left(\left(1 - \frac{2m}{r}\right)^{-\frac{1}{2}}, 0, 0, 0 \right).$$

The only non-zero component of the 4-acceleration is

$$a^1 = \frac{du^1}{d\tau} + \Gamma^1_{00} \left(u^0\right)^2$$

where

$$\Gamma^1_{00} = \frac{m}{r^2} \left(1 - \frac{2m}{r}\right).$$

As $u^1 = 0$ we obtain

$$a^1 = \frac{m}{r^2}, \qquad \text{and} \qquad a_1 = g_{11} a^1.$$

Finally ,

$$a^\mu a_\mu = - \left(\frac{m}{r^2}\right)^2 \left(1 - \frac{2m}{r}\right)^{-1},$$

so

$$a = \frac{m}{r^2} \left(1 - \frac{2m}{r}\right)^{-\frac{1}{2}}.$$

Note that the proper acceleration goes to infinity at the horizon, so only a photon can hover at the horizon.

2.8.2 Surface gravity

What is the gravitational acceleration at the event horizon as seen from infinity? The result in fact has the Newtonian form m/r^2. (Note that this is not the locally measured gravitational acceleration at the hole, which we have just calculated in the previous section, so it is not the same as Newtonian gravity, nor is it the Newtonian approximation at large distances that we are dealing with, even though this has the same form.) We shall give two ways of arriving at this answer, offering different levels of physical insight and mathematical sophistication. First a simple physical approach.

Consider a massless inextensible string used by a distant observer to raise a particle of unit mass at a radius r through a distance dl. In this process the local energy of the particle increases by $dE_r = g_r dl$, where g_r is the acceleration due to gravity at r. The energy expended by the distant observer in this process is $dE_\infty = g_\infty dl$. By conservation of energy the two energies are related by a redshift factor. (Otherwise one could make a perpetual motion machine by converting energy to radiation and sending it between the two ends of the string.) Thus

$$\frac{g_r}{g_\infty} = \frac{E_r}{E_\infty} = \left(1 - \frac{2m}{r}\right)^{-1/2} \qquad (2.36)$$

so

$$g_\infty = \frac{m}{r^2}. \qquad (2.37)$$

Here g_r is the proper acceleration felt by an observer hovering at radius r, which would be interpreted as the acceleration due to gravity at the radius r. It is given by (2.8.1). As $r \to 2m$, $g_r \to \infty$. Nevertheless, the force required at infinity to hold the particle of unit mass hovering at the horizon is $g_\infty = m/(2m)^2 = 1/4m$. We call this the *surface gravity* κ of the black hole. Our result is therefore that the surface gravity of a Schwarzschild black hole is $\kappa = 1/4m$.

2.8.3 Rindler coordinates

To investigate further what the spacetime is like in the vicinity of the event horizon we introduce a new set of coordinates in this region. For the time coordinate we use the Schwarzschild t, but we define a new radial coordinate ξ centred on $r = 2m$ and symmetrical about this origin by

$$r - 2m = \frac{\xi^2}{8m}$$

where the factor of 8 has been inserted with hindsight to tidy up the resulting expressions. We also define

$$\kappa = \frac{1}{4m},$$

which we have seen is the surface gravity of the black hole. Then we find

$$1 - \frac{2m}{r} = \frac{(\kappa\xi)^2}{(1 + \kappa^2\xi^2)} \qquad (2.38)$$

$$\sim \kappa^2\xi^2 \qquad (2.39)$$

near $\xi = 0$ (i.e. near $r = 2m$). So near the horizon the metric is approximately

$$d\tau^2 \sim \kappa^2\xi^2 dt^2 - d\xi^2 - \frac{1}{4\kappa^2}d\widetilde{\omega}^2, \qquad (2.40)$$

where

$$d\widetilde{\omega}^2 = d\theta^2 + \sin^2\theta d\phi^2.$$

The coordinates (t, ξ, θ, ϕ) are called Rindler coordinates.

This metric has a simple interpretation. Consider just the (t, ξ) plane and let

$$T = \xi \sinh \kappa t \qquad (2.41)$$
$$X = \xi \cosh \kappa t.$$

Substitution into the (t, ξ) part of the Rindler metric (2.40) shows that

$$d\tau^2 = dT^2 - dX^2, \qquad (2.42)$$

which is just Minkowski spacetime in two dimensions.

Problem 28 *Derive equation (2.42).*

So for an observer hovering near the horizon the Rindler metric is just Minkowski spacetime in an unusual coordinate system! Near the horizon of a spherical black hole the geometry is approximately flat.

This may require some explanation. To remain near the horizon keeping r close to $2m$ an observer must apply an acceleration. (Un-accelerated trajectories are radial geodesics which fall into the hole.) It is along this accelerated trajectory that the metric is approximately flat. This is a special feature. In a general spacetime it is only (unaccelerated, freely falling) geodesic observers who see their neighbourhoods as approximately flat.

Problem 29 *The proper time interval given by the Rindler metric in two dimensions is*

$$d\tau^2 = \kappa^2 \xi^2 dt^2 - d\xi^2.$$

On the curve $\xi = 1/a = constant$ we have, from Eq. (2.41),

$$T = a^{-1} \sinh \kappa t,$$
$$X = a^{-1} \cosh \kappa t.$$

From this calculate the components of the proper acceleration on this trajectory in (T, X) coordinates, namely $d^2T/d\tau^2$ and $d^2X/d\tau^2$, and show that the magnitude of the proper acceleration is a.

Thus a is the proper acceleration on the curve $\xi = 1/a$. From Eq. (2.39) $a \sim \kappa \left(1 - \frac{2m}{r}\right)^{-1/2}$. Comparing this with (2.8.1) we see that κ is approximately the acceleration of a particle near the horizon as seen from infinity, a relation that becomes exact at the surface of the hole at $r = 2m$. The quantity κ is therefore equivalent to the Newtonian acceleration due to gravity ("g") which again justifies calling it the surface gravity of the hole.

2.9   Other coordinates

The Schwarzschild metric appears to be a viable metric, and a solution of Einstein's equations, for $r < 2m$. However, in this region, the coordinate r is timelike and the coordinate t spacelike! For example, for a body in the region $r < 2m$ with $\theta = $ constant, $\phi = $ constant, $t = $ constant we have $d\tau^2 = g_{11}dr^2 > 0$ so a (non-zero) dr is a timelike interval in this region. This means that a body in this region cannot remain at constant r (since dr must be non-zero to make the worldline of the body timelike). We shall see that it must fall into the singularity at $r = 0$. Furthermore, since the metric coefficients depend on time, the metric within $r = 2m$ is not static.

However, the connection between the space within $r = 2m$ and outside $r = 2m$ cannot be examined in Schwarzschild coordinates, because the metric is invalid at $r = 2m$. We say that there is a coordinate singularity at $r = 2m$, because the problem turns out to be with the coordinate system, not with the physics. (A similar problem occurs at the pole of a spherical polar coordinate system in Euclidean space, where $\theta = 0$, $0 \leq \phi \leq 2\pi$ is not a circle, but a single point.) We can remedy this by finding one or more alternative sets of coordinates that do not have a coordinate singularity at $r = 2m$. We shall follow the historical development, introducing first the two sets of Eddington–Finkelstein coordinates and then the Kruskal coordinates, which cannot be expressed in elementary functions, but which have the advantage that they cover the whole spacetime.

2.9.1   Null coordinates

Consider first Minkowski spacetime. On an outgoing light ray (r increasing as t increases), $u = t - r$ is a constant. Different values of u label different light rays according to the initial point (the value of r when $t = 0$). As usual for convenience we use units with $c = 1$. Similarly, on ingoing rays the quantity $v = t + r$ is a constant, different values of the constant labeling different rays. Along an ingoing ray u changes, so u labels points on an ingoing ray. Conversely v labels points on a given outgoing ray. Because they label light rays the coordinates u and v are often called null coordinates. We can use either u or v (or both) as an alternative 'time' coordinate. For example, making the transformation

$$t = v - r \tag{2.43}$$

we get

$$d\tau^2 = dv^2 - 2dvdr - r^2d\widetilde{\omega}^2.$$

Using u and v as coordinates in the (r, t) plane, we have

$$d\tau^2 = dudv - \frac{1}{4}(v - u)^2d\widetilde{\omega}^2.$$

2.9.2 *Eddington–Finkelstein coordinates*

To carry out the analogous transformation in Schwarzschild spacetime we need to find the equation of a radial null ray. This is obtained by integrating $d\tau^2 = 0$ for a radial ray. We have from (2.8)

$$dt^2 = \frac{dr^2}{\left(1 - \frac{2m}{r}\right)^2} = dr_*^2, \tag{2.44}$$

where we have defined the new radial coordinate r_* in such a way that radial null rays are $d(t \pm r_*) = 0$ or $t \pm r_* = $ constant. Integrating the second pair of equations in Eq. (2.44) we get

$$r_* = r + 2m \log \left| \frac{r - 2m}{2m} \right|. \tag{2.45}$$

Let

$$v = t + r_*.$$

We are going to use v as a time coordinate, replacing t. Substitution in the Schwarzschild metric gives

$$d\tau^2 = \left(1 - \frac{2m}{r}\right)(dt^2 - dr_*^2) - r^2 d\widetilde{\omega}^2 \tag{2.46}$$

$$= \left(1 - \frac{2m}{r}\right) dv^2 - 2dr dv - r^2 d\widetilde{\omega}^2. \tag{2.47}$$

The final expression is the form of the Schwarzschild metric in ingoing Eddington-Finkelstein coordinates. In this form the metric is non-singular at $r = 2m$, because $\det(g) = -(2m)^4 \sin^2 \theta \neq 0$. To plot the spacetime in a (t, r) plane we introduce the new time coordinate corresponding to v and r,

$$t_* = v - r$$

(compare (2.43)). Figure 5 shows the (t_*, r) plane extending from $r = 0$ to $r = \infty$. In particular we can see what happens to the light cones as we approach $r = 2m$, which is a non-singular surface in these coordinates. Either from the figure, or from the metric (2.47), we see that $r = 2m, \theta = $ constant, $\phi = $ constant is a null ray, and that the surface $r = 2m$ is generated by these null rays (for different values of θ and ϕ).

2.10 Inside the black hole

We are now in a position to consider the properties of the black hole inside the event horizon. We can show that within the surface $r = 2m$ all future pointing directions point inwards to decreasing r. This means it is impossible to escape from within the region $r < 2m$, since real bodies can travel only in timelike (or null) directions.

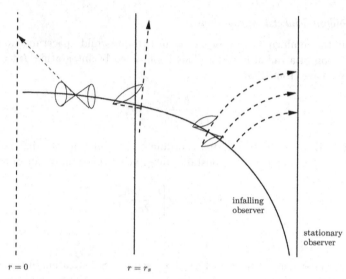

Figure 5 Spacetime diagram in Eddington–Finkelstein coordinates showing an observer falling towards a black hole.

The surface $r = 2m$ therefore acts as a one-way membrane for matter and radiation: nothing can escape into the surrounding spacetime. The surface $r = 2m$ is the event horizon, and the region $r < 2m$ is a 'black hole'.

To demonstrate this take the metric (2.47) and write it as

$$2drdv = -\left[d\tau^2 - \left(1 - \frac{2m}{r}\right)dv^2 + r^2d\widetilde{\omega}^2\right] \tag{2.48}$$

$$= -\left[d\tau^2 + \left(\frac{2m}{r} - 1\right)dv^2 + r^2d\widetilde{\omega}^2\right]. \tag{2.49}$$

For a timelike or null displacement $d\tau^2 \geq 0$ so for $r < 2m$ the sum of squares on the right hand side is positive and the overall expression is negative. Recall now that on a radial null ray $dv = d(t + r_*) = d(t_* + r) = 0$, and hence that a displacement dv lies within the future light cone if $dv > 0$. So to make the left hand side negative for a future-pointing displacement we have to have $dr < 0$. Thus a future-pointing displacement is necessarily in an inward direction.

The event horizon cuts off all events within $r = 2m$ from the outside world, so an explorer who crosses the event horizon of a black hole can never return or signal his experiences back to the outside world. A more pressing problem is that the observer cannot avoid destruction at the singularity that lies in their future.

2.10.1 The infalling observer

Our discussion of section 2.4.3 does not apply inside the black hole, because there are no stationary observers there. We can however transform to infalling observers outside the black hole and continue the results for these into the interior. This will enable us to provide a picture of what the infalling observer sees approaching the singularity.

Let the plunging observer fall from rest at infinity and denote his local frame of reference by S. As before, his velocity with respect to a stationary observer is

$$v = \left(\frac{2m}{r} \right)^{1/2}. \tag{2.50}$$

Let the stationary observer's local frame be S'. The angle of the incoming light ray in this frame is ψ'; let the corresponding angle in the S-frame be ψ. The usual special relativistic velocity transformation applies between the local frames, so

$$\cos \psi = \frac{\cos \psi' - v}{1 - v \cos \psi'} \tag{2.51}$$

(recalling that we use $c = 1$). Thus $\cos \psi < \cos \psi'$ so $\psi > \psi'$: the observer in S sees less of the sky as black.

Using now our previous result (2.30) for $\cos \psi'$ we have

$$\cos \psi = \frac{\left[1 - \left(1 - \frac{2m}{r} \right) \frac{b^2}{r^2} \right]^{1/2} - v}{1 - v \left[1 - \left(1 - \frac{2m}{r} \right) \frac{b^2}{r^2} \right]^{1/2}}. \tag{2.52}$$

What happens to (2.52) as $r \to 2m$? Inserting $r = 2m$ directly gives an indeterminate form. So we put $r = 2m + \delta$ and expand both numerator and denominator by the binomial theorem for small δ to get

$$\cos \psi \simeq -\frac{1 - \frac{b^2}{4m^2}}{1 + \frac{b^2}{4m^2}}.$$

The maximum value of b for which a ray can reach the horizon is again $\sqrt{27}m$, so inserting this value of b gives

$$\cos \psi_{\text{max}} = -\frac{23}{31}$$

or $\psi_{\text{max}} = 138°$. All rays up to this corresponding to smaller values of b can be seen.

Evidently the aberration formula (2.52) applies also inside the horizon. The angle ψ varies smoothly as the horizon is crossed. So let us see what the plunging observer sees at $r = m$ and as $r \to 0$. At $r = m$ (2.52) together with (2.50) and $b = \sqrt{27}m$ gives $\cos \psi_{\text{max}} = -0.598$, or $\psi_{\text{max}} = 127°$. Thus the observer sees more of the sky as black as he falls in.

As $r \to 0$ (2.52) gives

$$\cos \psi \sim -\left(\frac{r}{2m}\right)^{1/2} \tag{2.53}$$

independent of b. Hence in the limit all rays are seen at $\cos \psi = 0$, or $\psi = \pi/2$. As the observer crashes into the singularity the outside world appears as a thin bright ring.

Problem 30 *Derive (2.53) from equation (2.52).*

2.11   White holes

The Eddington–Finkelstein coordinates we used above were based on inward going null rays that entered $r = 2m$ in the future. It is natural to ask if we can make a similar picture with outgoing rays by taking $u = t - r_*$ as the 'time' coordinate. Straightforward substitution into the Schwarzschild metric gives

$$d\tau^2 = \left(1 - \frac{2m}{r}\right) du^2 + 2drdu - r^2 d\widetilde{\omega}^2.$$

This describes (part of) a spacetime which is the time-reverse of the black hole in which all future-directed null or timelike paths emerge from $r < 2m$ into the surrounding space. This object is called a white hole. Note that both black holes and white holes are mathematical solutions of the vacuum Einstein equations but that does not mean they are both physical possibilities. We shall present arguments later (chapter 6) to support the existence of black holes, but there is no evidence for the existence of white holes.

2.12   Kruskal coordinates

The Eddington–Finkelstein coordinates cover either the part of spacetime representing a black hole region or the part representing a white hole, but not both. It would obviously be convenient to be able to look at both regions in the same picture. This is made possible by yet another coordinate system, the Kruskal coordinates.

Return to the Schwarzschild metric in the form

$$d\tau^2 = \left(1 - \frac{2m}{r}\right)(dt^2 - dr_*^2) - r^2 d\widetilde{\omega}^2$$

and write this in double null coordinates u and v as

$$d\tau^2 = \left(1 - \frac{2m}{r}\right) dudv - r^2 d\widetilde{\omega}^2.$$

Now define

$$U = -\exp(-u/4m) \tag{2.54}$$
$$V = \exp(v/4m).$$

This can be shown to lead to

$$d\tau^2 = \frac{32m^3}{r}\exp(-r/2m)dU\,dV - r^2 d\widetilde{\omega}^2, \qquad (2.55)$$

where r is considered to be a function of U and V through

$$VU = -\exp\left(\frac{v-u}{4m}\right) = -\exp(r_*/2m),$$

since $v - u = 2r_*$. The Kruskal coordinates are (U, V, θ, ϕ). In Kruskal coordinates light rays are represented by lines at 45° (with $c = 1$).

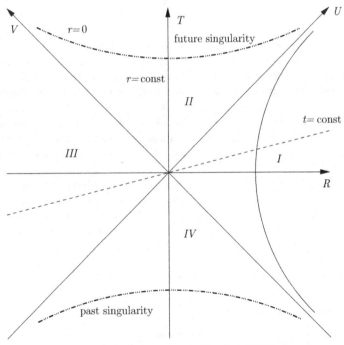

Figure 6 Schwarzschild spacetime in Kruskal (U, V) coordinates.

The metric (2.55) covers four regions as shown in Fig. 6. The two regions $r > 0$ and $r < 0$ are separate asymptotically flat universes. Note the two singularities at $r = 0$, one in the past and one in the future, which appeared separately in the pictures of the two sets of Eddington–Finkelstein coordinates. Figure 7b shows the surface of a collapsing star within which the metric would differ from the Kruskal picture.

There is one aspect of the Kruskal picture that can be misleading. Each point in the plane represents a 2-sphere of that radius at that time. This is not the same as saying that either of the angles can be re-created by rotating the diagram about an axis (which is how one often re-introduces an additional dimension). It is absolutely not the case that one can get from region I to region II, even in principle, by moving on a circle of constant ϕ! (And conversely, of course, rotation in Euclidean space in spherical polar coordinates does not take one to a region of negative r.)

2.12.1   The singularities at $r = 0$ and cosmic censorship

After falling through the event horizon a particle must continue to move to smaller values of r. This can continue in principle until the particle approaches $r = 0$ where the curvature becomes infinite. Any real body will suffer large tidal distortions as it approaches $r = 0$ and will be destroyed. But what happens to our hypothetical point particle? It reaches the edge of spacetime beyond which the theory has nothing to say. This is a true physical singularity, unlike the coordinate singularity at $r = 2m$.

To a certain extent it is a harmless singularity as far as physics in the outside world is concerned. This is because the unpredictability is hidden behind an event horizon and does not affect the black hole exterior. The white hole singularity at $r = 0$ in the past is a different matter. Here any unpredictability emerges into the surrounding spacetime making physics impossible. (Because, for example, any detection of a particle can be attributed to its emergence from the singularity rather than having to have a prior cause within the physical universe.) Thus the white hole singularity is 'naked'.

We shall see further examples of naked singularities later. The cosmic censorship hypothesis (put forward by Penrose) seeks to make naked singularities illegal as far as 'normal' physics is concerned by proposing that naked singularities cannot form from normal matter. There is no proof of this conjecture, but attempts to violate it have only helped to frame it more precisely. For the present, we observe that the formation of a black hole by the collapse of matter is not a time-symmetric process and leads to a black hole without the associated white hole.

2.12.2   The spacetime of a collapsing star

Figure 7 shows how the collapse of a sphere of matter to a singularity at $r = 0$ can be represented in (a) Eddington–Finkelstein coordinates and (b) Kruskal coordinates respectively. Note that within $r = 2m$ the Eddington–Finkelstein space and time coordinates (t, r_*) swap roles so the singularity at $r_* = r = 0$ occurs at a constant 'time' and is therefore spacelike, despite what the picture appears to show at first glance! This is clearly represented in the Kruskal picture. Note also that the white hole region has disappeared to be replaced by a solution of the field equations for the interior of the star. Although we make no attempt to find an interior solution, the pictures are drawn under the (correct) assumption that, prior to the final collapse, the interior will be a regular spacetime with a *decreasing* curvature as we go through the star towards $r = 0$.

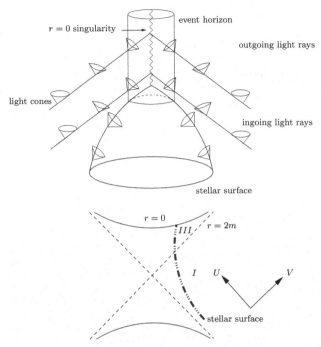

Figure 7   The spacetime of a collapsing star in (a) Eddington–Finkelstein coordinates and (b) Kruskal coordinates.

2.13   Embedding diagrams

The geometrical image of a cylinder as the surface of a tube is highly misleading in many respects. Intrinsically the cylinder has the geometry of flat space. The curvature one 'sees' when it is represented as a surface in three dimensional space comes from the embedding (the extrinsic geometry) of a 2-surface in three dimensional space. Nevertheless, the embedding picture of the cylinder does help to inform our feeling for what the geometry is like. To get some feel for the global geometry of the Schwarzschild black hole we can try to represent aspects of it by embeddings in three-space in a similar way.

To do this of course, we have to suppress all but two dimensions. Let us look at the spaces of constant time and also suppress one of the angular coordinates, by putting, say, $\theta = \pi/2$. The metric is

$$dl^2 = \left(1 - \frac{2m}{r}\right)^{-1} dr^2 + r^2 d\phi^2.$$

The metric of the Euclidean embedding space is

$$dl^2 = dz^2 + dr^2 + r^2 d\phi^2,$$

which on $z = z(r)$ becomes

$$dl^2 = (1 + z'^2)dr^2 + r^2 d\phi^2,$$

where $z' = dz/dr$. These are the same if

$$z = 2(2m)^{1/2}(r - 2m)^{1/2}. \tag{2.56}$$

This gives the surface in Fig. 8.

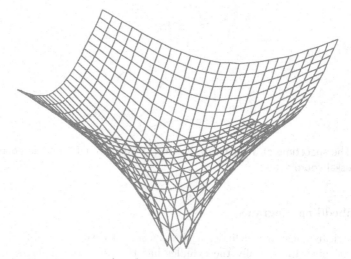

Figure 8   A plot of equation (2.56) showing the embedding surface with the same geometry as Schwarzschild in the $r - \phi$ plane.

This is the way in which black holes are customarily represented in popular literature. Since the Schwarzschild coordinates are not valid for $r = 2m$ we cannot follow the geometry further in by this means. If we look at the Kruskal picture however we see that through the bifurcation point at the origin the two Schwarzschild regions join together. For this cross-section (such as the one labelled $t = $ constant in Fig. 6 we get the Einstein-Rosen bridge as the embedding diagram, Fig. 9.

Although we cannot show it here, the only difference between this and a general section through the Kruskal metric, such as the one labelled $(U + V) = $ const. in Fig. 6, is that the throat region is extended and narrower, and breaks into two separate pieces if the section intersects $r = 0$.

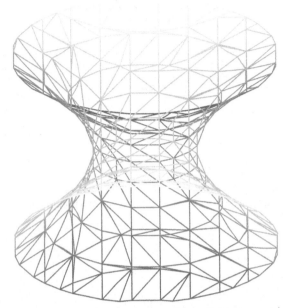

Figure 9   The Einstein–Rosen Bridge.

The two regions below and above the throat in Fig. 9 are two separate asymptotically flat universes. One might anticipate the possibility of travelling through the throat from one to the other. However, a glance at the full Kruskal picture shows that any attempt must end up in the singularity. In effect, the throat closes off faster than an observer can travel through it. On the other hand, it is not hard to imagine a picture in which the throat would connect two parts of the same Universe, in which case we have a *wormhole*. (See chapter 5.)

2.14   Asymptotic flatness

Far enough away from the central source of gravity the Schwarzschild spacetime approximates to Minkowski spacetime. Subject to certain technical definitions of what we mean by 'approximates to' such spacetimes are called asymptotically flat. For such spacetimes we can get what is sometimes a useful global picture of the causal structure (i.e. of which events can be causally connected) by using the Penrose–Carter diagrams. These make it possible to draw pictures of what is happening at infinity, by changing the coordinates to bring infinity to a finite coordinate value. This distortion is carried out in such a way that the relationship between light rays is maintained. We say that these diagrams show the causal structure of spacetimes (but clearly not their metric structure, since infinity is in the wrong place).

To see what this means consider the transformation from the Lorentzian coordinates (t, r, θ, ϕ) to new coordinates $(\psi, \xi, \theta, \phi)$ where

$$t = \frac{1}{2}\left(\tan\frac{1}{2}(\psi + \xi) + \tan\frac{1}{2}(\psi - \xi)\right) \qquad (2.57)$$

$$r = \frac{1}{2}\left(\tan\frac{1}{2}(\psi + \xi) - \tan\frac{1}{2}(\psi - \xi)\right).$$

The Minkowski metric becomes

$$d\tau^2 = \Omega^{-2}[d\psi^2 - d\xi^2 - \sin^2\xi(d\theta^2 + \sin^2\theta d\phi^2)] = \Omega^{-2}d\tilde{\tau}^2,$$

where $\Omega = 2\cos\frac{1}{2}(\psi + \xi)\cos\frac{1}{2}(\psi - \xi)$. Neglecting the overall factor Ω we get a new metric $d\tilde{\tau}^2$ which has the same light paths as $d\tau^2$ (because the vanishing of one implies and is implied by the vanishing of the other). The replacement of the metric $d\tau^2$ by the metric $d\tilde{\tau}^2$ is called a conformal transformation, with conformal factor Ω.

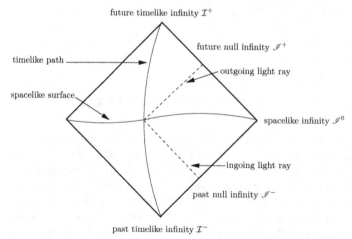

Figure 10   Asymptotic structure of Minkowski spacetime shown in a Penrose–Carter diagram.

The geometry of $d\tilde{\tau}^2$ is shown in Fig. 10. Light rays begin in the past at $\mathcal{I}^-$ and end at $\mathcal{I}^+$ (the lines $\psi + \xi = \pi$ and $\psi + \xi = -\pi$). These boundaries therefore represent light-like infinity. Causal relationships are respected by the transformations. Nevertheless, the paths of massive particles in the two metrics are not the same. All timelike paths in $d\tilde{\tau}^2$ start from I^- and end at I^+. Arranging that a finite range $(-\pi, \pi)$ of the new coordinates ψ and ξ cover an infinite range of the original (t, r) coordinates in (2.57), and making the conformal transformation, enables us to form a finite picture of an infinite spacetime. Such pictures are called Penrose or Penrose–Carter diagrams. In the next section we consider the Penrose–Carter diagram for Schwarzschild spacetime.

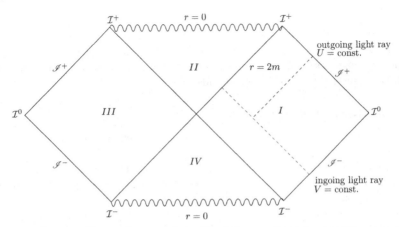

Figure 11 Penrose–Carter diagram of the Kruskal spacetime showing the asymptotic structure. Outgoing light rays end at null infinity on $\mathcal{I}^+$ and incoming rays begin at null infinity on $\mathcal{I}^-$. Spacelike infinity is denoted I^0 and timelike future and past infinity I^+ and $\bar{I}^-$ respectively. Light rays are given by $U = $ constant and $V = $ constant.

2.14.1 The Penrose–Carter diagram for the Schwarzschild metric

Our extension of the spacetime has taken us from the exterior region covered by the Schwarzschild coordinates to the 'whole' spacetime covered by the Kruskal coordinates. Although we have not proved it, in fact no further extensions are possible. We can put this together with a further conformal transformation that has no effect near the horizon, but which brings infinity to a finite distance. This provides us with a Penrose–Carter diagram of the black hole and allows us to picture its causal structure (Fig. 11).

The figure gives a view of the (U, V) plane.

2.14.2 The Penrose–Carter diagram for the Newtonian metric

Out of interest we can give a similar treatment for the Newtonian metric. The picture at infinity is the same as that for flat spacetime, but we need new coordinates to look at $r = 2m$. Confining ourselves to the (t, r) plane we put

$$dr = \pm \left(1 - \frac{2m}{r}\right)^{1/2} dR.$$

Then

$$d\tau^2 = \left(1 - \frac{2m}{r}\right)(dt^2 - dR^2).$$

The metric $d\bar{\tau}^2 = d\tau^2/(1 - 2m/r) = dt^2 - dR^2$ is therefore flat. Near $r = 2m$ we have

$$R \sim 2\sqrt{2m}(r - 2m)^{1/2},$$

so $r = 2m$ corresponds to $R = 0$. The metric is singular at $R = 0$, because $r < 2m$ clearly does not yield a Lorentzian metric. (Of course, near $r = 2m$ the Newtonian metric is not a physical gravitational field, because it is not even an approximate solution to Einstein's equations, but that does not stop us from considering the spacetime geometry to which is corresponds.)

2.15 Non-isolated black holes

There are several distinct concepts that coalesce in the Schwarzschild black hole, but which are different in general for black holes in a non-empty environment, for example for a black hole accreting material from the surrounding medium. In general we have to distinguish the following: the infinite redshift surface, the event horizon and the apparent horizon.

2.15.1 The infinite redshift surface

The surface defined by $g_{00} = 0$ has an infinite redshift to an external observer. To determine if a point belongs to the infinite redshift surface requires a knowledge of the metric just at that point, so this surface is a local construction. Thus it will not necessarily coincide with the event horizon in general.

2.15.2 Trapped surfaces

Consider a two dimensional spacelike surface and imagine that it emits a flash of light. A lightfront moves normal to the surface inwards and outwards. Under normal circumstances the area of the inward moving lightfront decreases and that of the outward moving one increases. Equivalently, the inward normals to the inward moving light front converge and the outward normals to the outward moving front diverge.

However, if we repeat this experiment inside a black hole, both sets of normals converge (Fig. 12). Our two dimensional spacelike surface is said to be trapped.

2.15.3 Apparent horizon

The outermost trapped surface is the apparent horizon. In the vacuum Schwarzschild solution this is the surface $r = 2m$. However, the apparent event horizon is clearly a local construction (you can measure the convergence or divergence locally), whereas the ('true') event horizon is a global property: it depends on constructing null geodesics to determine whether they reach infinity or not, which cannot be judged from the local behaviour.

For a concrete example consider a black hole accreting a spherical shell of matter (Frolov and Novikov, 1998). A light ray which was heading to infinity before the infall of the shell is subject to a stronger gravitational field after the matter is accreted, so may not in fact escape to infinity. Thus the original ray is part of the event horizon even though it was not initially trapped. The apparent horizon

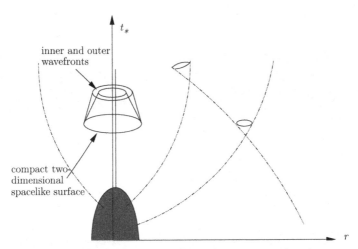

Figure 12   Both outgoing and ingoing wavefronts from a trapped surface converge. The trapped surface is inside a black hole formed by collapse.

therefore lies inside the event horizon. One can show that this is a general feature of the two horizons.

2.16   The membrane paradigm

Although the event horizon is not a physical boundary (and is locally undetectable) it is intuitively convenient to think of a black hole as a physical object, endowed with physical properties that encode its relation to the external world. To do this rigorously, we imagine the black hole surrounded by a membrane to which are attributed appropriate physical properties. This is the so-called 'membrane paradigm' (Thorne *et al.*, 1986). For example, to express the influence of a black hole on external electromagnetic fields the membrane is endowed with a surface resistivity allowing charges and currents to terminate electric and magnetic field lines. Similarly, to account for the deformations of a black hole in the presence of, for example, an orbiting planet, the membrane can be given certain elastic properties. Normal physical bodies have timelike surfaces. The membrane is therefore chosen not to be coincident with the horizon, which is a null surface, but as a timelike surface just outside the horizon. The black hole can then be thought of from the outside as a normal body in spacetime. This approach works not only for Schwarzschild black holes but also for the more general vacuum black holes of the next chapter, and also for black holes in general, which we do not consider here. The details are beyond the scope of this book but can be found in Thorne *et al.* (1986).

Chapter 3

ROTATING BLACK HOLES

We saw in the previous chapter that the Schwarzschild metric is the only spherically symmetric vacuum solution of Einstein's field equations and that this represented the unique end-point of the collapse of a spherical star. Since stars are not strictly spherical but axisymmetric about an axis of rotation, it is natural to seek axially symmetric vacuum solutions of Einstein's equations to represent the spacetime outside a rotating star. Such solutions can be found and may be of interest, for example in the structure of rotating neutron stars. For a slowly rotating body of mass M with angular momentum J the Einstein equations outside the body are solved approximately by the metric

$$ds^2 \approx \left(1 - \frac{2GM}{c^2 r}\right) c^2 dt^2 + \frac{4GJ \sin^2 \theta}{c^2 r} dt d\phi - \left(1 + \frac{2GM}{c^2 r}\right) dr^2 - r^2 d\theta^2$$
$$- r^2 \sin^2 \theta d\phi^2, \tag{3.1}$$

to order $1/r$. As $J \to 0$ we recover the Schwarzschild solution.

Another way of approaching this question is to imagine a star acquiring angular momentum by capture of infalling, orbiting material. Outside the star the metric will be given by (3.1). Now imagine the capture of an infalling particle from the last stable orbit by a Schwarzschild black hole. We have already seen that all but 5.7 per cent of the rest mass energy of the particle is added to the mass of the black hole. In addition, the black hole captures the angular momentum of the matter in the last stable orbit. This leads to a spinning black hole. One might imagine again that the metric would be given by (3.1). Far from a spinning black hole this is true; close to a spinning black hole it is not even approximately the case (because the $O(1/r^2)$ terms cannot be neglected), no matter how slowly the black hole is rotating. Instead we have a new solution, the Kerr solution, which gives the pure vacuum solution corresponding to a spinning black hole. Note that, conversely, the Kerr solution does not appear to correspond to the field outside a rotating material body, since no interior solutions with physical matter have been found that fit smoothly on to a Kerr exterior metric.

We shall next give the Kerr solution without proof. (Even to check by direct substitution that it satisfies the vacuum Einstein equations is a major undertaking!)

3.1   The Kerr metric

The following metric is a solution of the Einstein field equations in a vacuum:

$$ds^2 = \frac{(\Delta - a^2 \sin^2 \theta)}{\rho^2} c^2 dt^2 + \frac{4GMa}{c\rho^2} r \sin^2 \theta d\phi dt - \frac{\rho^2}{\Delta} dr^2 - \rho^2 d\theta^2 - \frac{A \sin^2 \theta}{\rho^2} d\phi^2,$$

where

$$\Delta = r^2 - \frac{2GM}{c^2} r + a^2, \tag{3.2}$$

$$\rho^2 = r^2 + a^2 \cos^2 \theta, \tag{3.3}$$

$$A = (r^2 + a^2)^2 - a^2 \Delta \sin^2 \theta, \tag{3.4}$$

which we shall find to be the metric of a black hole of mass M and angular momentum $J = aMc$. This solution is the Kerr metric in Boyer-Lindquist coordinates (t, r, θ, ϕ). In these coordinates the metric is approximately Lorentzian at infinity (i.e. it is Minkowski space in the usual coordinates of special relativity). For $a = 0$ we recover the Schwarzschild solution in Schwarzschild coordinates.

Note that the parameter a has units of length. The metric can be simplified by defining time and masses in length units also. To do this we put: $m = GM/c^2$ and replace $s = c\tau$ by τ and ct by t as we did for the Schwarzschild metric. The result is equivalent to setting $G = 1$ and $c = 1$. We obtain

$$d\tau^2 = \frac{(\Delta - a^2 \sin^2 \theta)}{\rho^2} dt^2 + \frac{4ma}{\rho^2} r \sin^2 \theta d\phi dt - \frac{\rho^2}{\Delta} dr^2 - \rho^2 d\theta^2 - \frac{A \sin^2 \theta}{\rho^2} d\phi^2. \tag{3.5}$$

or, with a slight rearrangement, in an equivalent form

$$d\tau^2 = \frac{\Delta}{\rho^2} (dt - a \sin^2 \theta d\phi)^2 - \frac{\sin^2 \theta}{\rho^2} [(r^2 + a^2) d\phi - a dt]^2 - \frac{\rho^2}{\Delta} dr^2 - \rho^2 d\theta^2, \tag{3.6}$$

where

$$\Delta = r^2 - 2mr + a^2, \tag{3.7}$$

and ρ and A are unchanged. For future reference we define the angular momentum of the black hole $J = aMc$ in physical units. In geometrical units we put $j = am = GJ/c^3$. The units of j are (metres)^2.

3.2   The event horizon

The first property of the Kerr metric that we notice is that as $\Delta \to 0$ the metric coefficient $g_{11} \to \infty$. So this appears to be the analogue of the Schwarzschild condition $r \to 2m$. Indeed, if $a = 0$, then $\Delta = 0$ when $r = 2m$. Anticipating somewhat, we see that $\Delta = r^2 + a^2 - 2mr = 0$ is a quadratic in r and so has two solutions $r_\pm$. It is indeed the case that the outer of these solutions,

$$r_+ = m + \sqrt{m^2 - a^2}, \tag{3.8}$$

is the location of the event horizon in the Kerr metric. The justification that r_+ is the event horizon in the equatorial plane will be provided in section 3.10, and that it is the event horizon in general will be apparent from the Penrose–Carter diagram in section 3.18. The inner solution r_- has a different physical significance which we shall explain in section 3.18. Here we shall evaluate two quantities associated with the event horizon to provide some practice with the Kerr metric and for future reference.

3.2.1 The circumference of the event horizon

The (proper) circumference of the curve $r =$ constant in the equatorial plane $\theta = \pi/2$ at constant Boyer-Lindquist time t, is

$$\int_0^{2\pi} (g_{33})^{1/2} d\phi = \int_0^{2\pi} \frac{[(r^2 + a^2)^2 - a^2\Delta]^{1/2}}{r} d\phi.$$

At the event horizon, $r = r_+$, this is

$$2\pi \frac{(r_+^2 + a^2)}{r_+} = 4\pi m,$$

which is independent of a and equal to its value in the Schwarzschild case.

Although the horizon is a surface of constant radial coordinate the proper distance around the equator is not the same as that of a circle going through the poles which is given by

$$\int_0^{2\pi} \rho^2 d\theta.$$

Problem 31 *For the extreme Kerr black hole ($a = m$) show that the circumference of the event horizon in the equatorial plane is greater than the polar circumference and evaluate the latter approximately. (You will need to approximate an integral or evaluate it numerically.)*

3.2.2 The area of the event horizon

An important result in the classical theory of black holes is that the area of an event horizon cannot decrease, for example during a process of merger or of energy extraction. We shall defer a proof of this to the next chapter. Here we obtain an expression for the area A_h of the event horizon from the metric. The radius of the event horizon is $r_+ = m + (m^2 - a^2)^{1/2}$, and so, directly from Eq. (3.4), $A = (r^2 + a^2)^2 - a^2\Delta\sin^2\theta = 4m^2[m + (m^2 - a^2)^{1/2}]^2$ on the horizon. The area A_h is therefore

$$Area = A^{1/2} \int_0^{2\pi} d\phi \int_{-\pi/2}^{\pi/2} d\theta \sin\theta,$$

or,

$$A_h = 8\pi m[m + (m^2 - a^2)^{1/2}] \tag{3.9}$$

(since the angular integral is just the area of the unit sphere, or 4π).

Problem 32 *Putting $a = 0$ in the metric Eq. (3.5) we have seen that we recover the Schwarzschild metric for a mass m. But what do we get if we put $m = 0$ keeping $a \neq 0$? By making the following coordinate transformation on the Kerr metric with $m = 0$,*

$$x = \sqrt{r^2 + a^2}\sin\theta\cos\phi, \; y = \sqrt{r^2 + a^2}\sin\theta\sin\phi, \; z = r\cos\theta,$$

show that we recover flat spacetime.

3.3 Properties of the Kerr metric coefficients

In this section we look at some of the basic properties of the metric coefficients in Boyer-Lindquist coordinates. We shall see that the fact that the Kerr metric is not diagonal makes it somewhat more complicated to work with than the Schwarzschild metric.

3.3.1 Identities

By inspection of equation (3.5) we see that the metric coefficients take the following form:

$$g_{00} = \frac{(\Delta - a^2\sin^2\theta)}{\rho^2}, \tag{3.10}$$

$$g_{30} = g_{03} = \frac{2mar}{\rho^2}\sin^2\theta,$$

$$g_{11} = -\frac{\rho^2}{\Delta}, \quad g_{22} = -\rho^2, \quad g_{33} = -\frac{A\sin^2\theta}{\rho^2}.$$

The metric coefficients satisfy the following useful identities:

$$g_{33} + a\sin^2\theta g_{03} = -(r^2 + a^2)\sin^2\theta$$
$$g_{03} + a\sin^2\theta g_{00} = a\sin^2\theta$$
$$ag_{33} + (r^2 + a^2)g_{03} = -\Delta a\sin^2\theta$$
$$ag_{03} + (r^2 + a^2)g_{00} = \Delta$$
$$(g_{03})^2 - g_{00}g_{33} = \Delta\sin^2\theta.$$

All of these are proved by substitution from (3.10) and the definitions of Δ, ρ and A.

Problem 33 *Verify the last of these identities.*

3.3.2 Contravariant components

Since the Kerr metric is not diagonal, on account of the non-zero g_{03} ($= g_{30}$) term, the inverse metric cannot be found by taking the inverses of each of the components. To get the contravariant components of the metric tensor it is instead necessary to take the correct matrix inverse of $(g_{\mu\nu})$. This is most easily achieved by considering $(g_{\mu\nu})$ as a diagonal 2×2 block matrix of 2×2 matrices (with row and column order (t, ϕ, r, θ)) and taking the inverse of the diagonal blocks. The result for the non-zero entries is

$$g^{00} = \frac{(r^2 + a^2)^2 - a^2 \Delta \sin^2 \theta}{\rho^2 \Delta},$$

$$g^{03} = \frac{2mar}{\rho^2 \Delta} = g^{30},$$

$$g^{11} = -\frac{\Delta}{\rho^2},$$

$$g^{22} = -\frac{1}{\rho^2},$$

$$g^{33} = -\frac{\Delta - a^2 \sin^2 \theta}{\Delta \rho^2 \sin^2 \theta}.$$

These satisfy the useful identities

$$(g^{03})^2 - g^{00} g^{33} = \frac{1}{\Delta \sin^2 \theta},$$

$$ag^{00} - (r^2 + a^2)g^{03} = a,$$

$$(r^2 + a^2)g^{33} - ag^{03} = -\frac{1}{\sin^2 \theta}.$$

Furthermore, from

$$g_{\mu\sigma} g^{\mu\tau} = \delta^\tau_\sigma,$$

with $\sigma = 0$, $\tau = 3$ we deduce

$$\frac{g_{30}}{g_{00}} = -\frac{g^{03}}{g^{33}}.$$

Similarly, with $\sigma = 3$ and $\tau = 0$ we get

$$\frac{g^{03}}{g^{00}} = -\frac{g_{03}}{g_{33}}.$$

3.4 Interpretation of *m*, *a* and geometric units

As we have already seen, the Kerr metric (3.5) reduces to the Schwarzschild metric as $a \to 0$. To be consistent we expect that m must be the mass of the source of the gravitational field as measured by a distant orbiting body in geometric units (such that $G = 1$ and $c = 1$). The mass in conventional units would be c^2m/G. (But once again, note this is not the mass of an extended material object in the system, because the Kerr solution does not correspond to the metric outside a material body.)

As $r \to +\infty$ the Kerr metric (3.5) becomes

$$d\tau^2 = \left(1 - \frac{2m}{r}\right) dt^2 + \frac{4am\sin^2\theta}{r} dt d\phi - \left(1 + \frac{2m}{r}\right) dr^2 - r^2 d\theta^2 - r^2 \sin^2\theta d\phi^2. \quad (3.11)$$

This looks like the weak field of a slowly rotating body if we put $a = j/m$. Thus, from the point of view of an orbiting body at a large distance, the parameter a is the angular momentum per unit mass in geometric units, or ca is the angular momentum per unit mass in conventional units. We shall find other strong field effects that confirm this interpretation of a.

Problem 34 *Derive Eq. (3.11) from the Kerr metric.*

3.5 Extreme Kerr black hole

We shall find that the properties of the Kerr solution differ markedly according as $a < m$, $a = m$ or $a > m$. In fact, $a > m$ is unphysical for various reasons (section 3.18), in recognition of which the case $a = m$ is referred to as an *extreme* (or *extremal*, or *maximal*) Kerr black hole.

Problem 35 *By looking at the expression for the radius of the event horizon (3.8) show that a Kerr black hole with $a > m$ has no event horizon.*

3.6 Robinson's theorem

Just as the Schwarzschild metric is the unique static vacuum solution of Einstein's equations (Israel's theorem) so the Kerr metric is the unique stationary axisymmetric vacuum solution. This result is sometimes called Robinson's theorem or the Carter-Robinson theorem. Strictly both of these results require some further technical assumptions that amount to requiring the spacetime to be asymptotically flat and non-singular in the exterior of an event horizon. If the spacetime is taken to be that of a black hole, then the assumption of axisymmetry can be dropped: stationary black holes must be axisymmetric.

The uniqueness theorems can be extended to include electromagnetic fields in the environment, hence to charged black holes and (at least in theory) black holes with magnetic monopole charges, so the most general black hole is defined by four scalar quantities: mass, angular momentum and electric and magnetic charges. The fact that there are no long range multipole fields attached to an isolated black hole is sometimes expressed by saying that 'black holes have no hair'.

3.7 Particle orbits in the Kerr geometry

In the following sections we explore the spacetime of the Kerr black hole by investigating the orbits of particles and the trajectories of light rays in the exterior of the hole. Just as in the Schwarzschild case we are interested in the behaviour of geodesics, which are the paths of particles falling freely under gravity. We use a similar short-cut based on the symmetries of the metric to obtain first integrals of the motion. From these we can find the motion under a variety of initial conditions. The results are quite different from any Newtonian analogue. A rotating Newtonian body adopts a non-spherical shape as a result of its rotation. Its gravitational field differs from that of a non-rotating body only because of this different shape. A non-rotating body of the same shape would have the same Newtonian gravitational field. A rotating relativistic body influences the surrounding matter in addition directly through its rotation. Near a rotating black hole this leads to some unexpected effects.

3.7.1 Constants of the motion

Consider a freely falling test particle with 4-velocity u^μ in the exterior of a Kerr black hole. Just as for the Schwarzschild solution (section 2.3.3) the covariant component of 4-velocity in a direction of symmetry is a constant. For the Kerr metric the symmetry directions in Boyer-Lindquist coordinates are $(k^\mu) = (1, 0, 0, 0)$ and $(k^\mu) = (0, 0, 0, 1)$. Thus, with E and L_z constants (and $c = 1$)

$$u_0 = E, \tag{3.12}$$
$$u_3 = -L_z.$$

The parameter E is the energy of the particle per unit mass, and L_z is the component of angular momentum of the particle, per unit mass, normal to the equatorial plane. Note that we choose u_3 to be negative to correspond to motion in the direction of increasing ϕ. This follows because

$$u^3 = \frac{d\phi}{d\tau} = g^{30}u_0 + g^{33}u_3$$

is positive for all r if u_3 is negative (since g^{33} is negative and the other terms are each positive). Let us emphasise again that E and L_z are the energy and angular momentum per unit mass measured by a distant stationary observer. A local observer will measure a different value for the energy of a particle as we showed for the Schwarzschild case in section 2.3.8. We should stress that L_z is the angular momentum per unit mass in the z-direction and not the total angular momentum as in the Schwarzschild case. In the Schwarzschild metric particles move in a plane and all planes are equivalent, so we were free to orient the coordinate system in order that the plane is always the equatorial plane and the angular momentum is the total. This enabled us to ignore the θ-geodesic equation. In the Kerr metric there is less symmetry and the equatorial plane is special. Thus when we interpret the constant in the limit of large r we get

$$u_3 = g_{33}u^3 = -r^2 \sin^2\theta d\phi/d\tau = -L_z,$$

and only for orbits in the equatorial plane is L_z the total angular momentum per unit mass.

There is a further constant of the motion related to the total angular momentum that is referred to as the Carter integral, Q. It can be shown that

$$u_2 = \pm \left\{ Q - \cos^2\theta \left[a^2(1 - E^2) + \frac{L_z^2}{\sin^2\theta} \right] \right\}^{1/2}. \tag{3.13}$$

The Carter integral Q must be zero for motion in the equatorial plane. This follows because the axial symmetry picks out the equatorial plane as special, so a particle moving in the equatorial plane remains there i.e. if $\theta = \pi/2$, then $d\theta/d\tau = u^2 = u_2 = 0$. There is no reason why other orbits should be confined to planes, and indeed they are not as we show in section 3.7.4. We shall see that orbits out of the equatorial plane are dragged round by the rotation of the hole.

Finally, the components of 4-velocity are related by

$$g^{\mu\nu}u_\mu u_\nu = 1$$

as usual (with $c = 1$).

3.7.2   Energy

Recall that we obtain the equations of motion of a test particle in the following way. We know that certain covariant components of the 4-velocity are constant; on the other hand it is the contravariant components that are related to the coordinate differentials. We therefore use the metric to connect the two sets of components. For the energy equation we have

$$\frac{dt}{d\tau} = u^0 = g^{00}u_0 + g^{03}u_3.$$

By substituting for the metric coefficients from (3.5) and using (3.12), we get

$$\frac{dt}{d\tau} = \frac{AE}{\rho^2\Delta} - \frac{2mar}{\rho^2\Delta}L_z \tag{3.14}$$

or

$$\rho^2\frac{dt}{d\tau} = a(L_z - aE\sin^2\theta) + \frac{r^2 + a^2}{\Delta}[E(r^2 + a^2) - aL_z], \tag{3.15}$$

where L_z is positive for rotation of the particle in the same sense as the hole and negative in the opposite sense. This is the equation for the conservation of relativistic energy in contravariant form.

3.7.3 Angular momentum

For the angular momentum equation we proceed similarly. We have

$$\frac{d\phi}{d\tau} = u^3 = g^{30}u_0 + g^{33}u_3,$$

from which

$$\rho^2 \frac{d\phi}{d\tau} = \frac{L_z}{\sin^2\theta} - aE + \frac{a}{\Delta}[E(r^2 + a^2) - L_z a], \tag{3.16}$$

where L_z is positive for co-rotation and negative for contra-rotation. This is the equation for the conservation of relativistic angular momentum in contravariant form. For motion confined to the equatorial plane this becomes

$$\Delta \frac{d\phi}{d\tau} = \frac{2ma}{r}E + L\left(1 - \frac{2m}{r}\right). \tag{3.17}$$

3.7.4 The Carter integral

The Carter integral gives the third equation of motion (replacing the condition that motion is in a plane in Schwarzschild). We have

$$u_2 = g_{22}u^2 = -\rho^2 \frac{d\theta}{d\tau},$$

which, with (3.13) gives

$$\rho^2 \frac{d\theta}{d\tau} = \pm\left\{Q - \cos^2\theta\left[a^2(1 - E^2) + \frac{L_z^2}{\sin^2\theta}\right]\right\}^{1/2}. \tag{3.18}$$

As we have seen, for $Q = 0$ there is a solution with $\theta = \pi/2$ for any E and L_z, which corresponds to motion in the equatorial plane. No other orbits can lie in a fixed plane. Orbits out of the equatorial plane are dragged round. For large r these orbits are approximately plane but have normals that precess slowly around the spin axis of the hole. (See section 3.14.) A simple deduction from Eq. (3.18) is that for $E \leq 1$, Q must be positive (or zero) for a solution to exist. (The square root must be real.) We shall see later that $E \leq 1$ corresponds to bounded motion. These orbits move between maximum and minimum values of θ (two turning points at which $d\theta/d\tau = 0$). Also, orbits for $Q = 0$ and $L_z = 0$ exist for $E \geq 1$. These correspond to a particle coming in from infinity with zero total angular momentum. Unbound orbits with $E > 1$ can have negative Q.

We can get some insight into the physical interpretation of the Carter integral Q by considering the motion of a particle at a large distance from the black hole (Frolov and Novikov, 1998). At large r the Kerr metric simplifies to the form given in equation (3.5). Since the angular part is the same as that of flat space, the total angular momentum per unit mass is

$$L^2 = r^4 \left[\left(\frac{d\theta}{d\tau} \right)^2 + \sin^2\theta \left(\frac{d\phi}{d\tau} \right)^2 \right]. \tag{3.19}$$

Now substitute for $d\theta/d\tau$ from equation (3.18) and for $d\phi/d\tau$ from equation (3.16). At large r these two equations become

$$r^2 \frac{d\phi}{\tau} = \frac{L_z}{\sin^2\theta}$$

and

$$r^2 \frac{d\theta}{d\tau} = \pm \left[Q - a^2\cos^2\theta \left(1 - E^2 \right) - L_z^2 \cot^2\theta \right]^{1/2}. \tag{3.20}$$

Substituting these expressions into (3.19) gives

$$L^2 = Q + L_z^2 + a^2 \cos^2\theta \left(1 - E^2 \right).$$

It is clear that the total angular momentum of the particle is not conserved and varies with θ. But the total angular momentum of the hole plus the particle must be conserved, so there is an exchange of angular momentum between the two. As $a \to 0$ this becomes

$$L^2 = Q + L_z^2. \tag{3.21}$$

Problem 36 *From Eq. (3.18) show that for bound orbits there are two (real) turning points for θ.*

3.7.5   The radial equation

The radial equation is most easily obtained from the scalar product $g^{\mu\nu}u_\mu u_\nu = 1$. Writing the equation in this form (with the contravariant metric and covariant components of velocity) simplifies the algebra because we can use the conserved quantities (3.12) and (3.13) directly. A few lines of algebra give

$$\rho^4 \left(\frac{dr}{d\tau} \right)^2 = [E(r^2 + a^2) - aL_z]^2 - \Delta[r^2 + (L_z - aE)^2 + Q]. \tag{3.22}$$

This is the first integral of the radial equation of motion. An alternative form as a quadratic in E is sometimes useful:

$$\rho^4 \left(\frac{dr}{d\tau} \right)^2 = r^2 \left(r^2 + a^2 + \frac{2ma^2}{r} \right) E^2 - 4mar EL_z - r^2 \left(1 - \frac{2m}{r} \right) L_z^2 - \Delta \left(Q + r^2 \right). \tag{3.23}$$

3.7.6   The effective potential

In the Schwarzschild case we found it useful to express the radial equation of motion in terms of an effective potential. Obtaining the analogous effective potential from the radial equation for the Kerr case is both a little tricky and not so useful. Nevertheless we shall obtain it for completeness and because it appears in the literature.

Looking back at the Schwarzschild case we see that at the turning points $dr/d\tau = 0$, we have $V_{\text{eff}}^2 = E^2$. To be consistent, and to have an effective potential that determines the possible motions in Kerr in a similar way, we now *define* the effective potential by this condition. Thus, we require $E = V_{\text{eff}}$ when the right side of (3.23) vanishes. Thus to find the effective potential we find the value of E that makes the right side of (3.23) zero and set this equal to V_{eff}. The quadratic in E is solved by

$$V_{\text{eff}} = \frac{2amL_z r \pm \sqrt{4a^2m^2L_z^2 r^2 - [r^4 + a^2(r^2 + 2mr)][L_z^2 a^2 - (r^2 + L_z^2 + Q)\Delta]}}{r^4 + a^2(r^2 + 2mr)}. \tag{3.24}$$

One surprise is that there are two effective potentials, depending on whether we add or subtract the square root term; call these V_+ and V_- respectively. Equation (3.22) in terms of the effective potentials is therefore

$$\rho^4 \left(\frac{dr}{d\tau}\right)^2 = [(r^2 + a^2)^2 - a^2\Delta](E - V_+)(E - V_-). \tag{3.25}$$

Motion is possible only where $dr/d\tau$ is real, which implies that either $E \geq V_+$ or $E \leq V_-$. In Schwarzschild we see that $V_- = -[(1 - 2m/r)(1 + L^2/r^2)]^{1/2} < 0$. The region $E \leq V_-$ is therefore excluded by the condition that the particle energy is positive in the exterior region of the black hole and only the one effective potential is relevant. By contrast we shall see that negative energies are allowed in the Kerr geometry.

A second surprise is just how complicated it is going to be to extract the general properties of motions in the Kerr geometry from the Kerr effective potential. We shall therefore confine ourselves here to some general comments and look at particular (and particularly illuminating) examples in the following sections.

Note first that $V_+ \to 1$ as $r \to \infty$. (There is a product of terms in r^4, one explicit and the other coming from $r^2\Delta$, that dominates the square root and cancel with the dominant r^4 in the denominator.) Hence for unbounded motion $E \geq 1$ (since $E \geq V_+$). In physical terms this says that the particle energy exceeds its rest mass energy, as we might expect. In fact, almost all orbits with $E \geq 1$ are unbounded in the sense that they escape to infinity. Conversely for $E < 1$ the orbits are always bounded (i.e. the particle cannot escape to infinity).

3.8   Frame-dragging

We now introduce a surprising, and at first sight contradictory feature of rotating black holes. Consider a particle falling from infinity with zero angular momentum, $L_z = 0$. The energy equation (3.15) simplifies to

$$\rho^2 \frac{dt}{d\tau} = \frac{[(r^2 + a^2)^2 - a^2\Delta\sin^2\theta]}{\Delta}E \tag{3.26}$$

and the angular momentum equation (3.16) to

$$\rho^2 \frac{d\phi}{d\tau} = \frac{2mar}{\Delta}E. \tag{3.27}$$

Despite the fact that initially the particle falls radially with no angular momentum, it acquires an angular motion during infall. The angular velocity as seen by a distant observer is

$$\omega(r,\theta) = \frac{d\phi}{dt} = \frac{d\phi/d\tau}{dt/d\tau} = \frac{2mar}{(r^2 + a^2)^2 - a^2\Delta\sin^2\theta}, \tag{3.28}$$

which is clearly non-zero at finite radii. Note that positive a implies positive ω, so the particle acquires an angular velocity in the direction of the spin of the hole. The angular momentum remains zero (as it must, since angular momentum is conserved). The distinction between angular momentum and angular speed arises here because of the non-zero g^{03} term in the metric. If $g^{03} = 0$ (as in Schwarzschild) then $u^3 = g^{33}u_3$, so $L_z = u_3 = 0$ implies that $u^3 = 0$ i.e. $d\phi/d\tau = 0$; in this case zero angular momentum is equivalent to zero angular speed. However if $g^{03} \neq 0$ then $u_3 = 0$ does not imply that $u^3 = 0$ (because $u^3 = g^{30}u_0 + g^{33}u_3$) and zero angular momentum is not the same as zero angular speed.

From a physical point of view we can interpret this phenomenon as a dragging round of the local inertial frames of reference by the rotating hole. An observer in a local inertial frame must find that the angular momentum of a body subject to no non-gravitational forces is conserved. We have just seen that such a frame rotates with an angular velocity ω relative to infinity, hence is dragged round with the rotation of the hole. The frame dragging diminishes towards the poles ($\theta = 0, \pi$).

At $\Delta = 0$, and $r = r_+$, the event horizon, we get for the angular velocity

$$\omega_+ = \frac{a}{2mr_+} = \omega_H \tag{3.29}$$

irrespective of the angle θ. This is called the angular velocity of the black hole. It is the angular speed with which local inertial frames of reference are dragged round on the surface of the horizon.

Problem 37 *From the relation*

$$\omega = \frac{d\phi}{dt} = \frac{u^3}{u^0},$$

and using the final identity from section 3.3.2 show that

$$\omega = \frac{g^{30}}{g^{00}} = -\frac{g_{03}}{g_{33}}.$$

We shall need this result later in section 3.9.

Problem 38 *Show that an observer at infinity attributes a speed (circumference/period) of c to a point on the equator of an extreme Kerr black hole.*

Problem 39 *Show that the angular speed of a slowly rotating hole (a $\ll$ m) is*

$$\omega_+ \approx \frac{a}{4m^2}.$$

The angular momentum is $j = ma$, and the moment of inertia is given, as usual, by $j = I\omega_+$. For a slowly rotating hole show that $I \approx mr_+^2$.

3.8.1 Free fall with zero angular momentum

We now consider the orbit of a particle having zero angular momentum falling inwards towards the horizon. Setting $Q = 0$ and $L_z = 0$ equations (3.14), (3.16), (3.18) and (3.22) become

$$\frac{dt}{d\tau} = \frac{AE}{\rho^2 \Delta} \tag{3.30}$$

$$\frac{d\phi}{d\tau} = \frac{2marE}{\rho^2 \Delta} \tag{3.31}$$

$$\frac{d\theta}{d\tau} = \pm \frac{[a^2 (E^2 - 1) \cos^2 \theta]^{\frac{1}{2}}}{\rho^2} \tag{3.32}$$

$$\frac{dr}{d\tau} = \pm \frac{\left[E^2 (r^2 + a^2)^2 - \Delta (r^2 + a^2 E^2) \right]^{\frac{1}{2}}}{\rho^2}. \tag{3.33}$$

Now from (3.30) and (3.33) a distant observer sees $\frac{dr}{dt} = \frac{dr/d\tau}{dt/d\tau} \to 0$ as the particle approaches the horizon at $\Delta = 0$. From (3.30) and (3.31) $\frac{d\phi}{dt} \to \omega_+$. In other words, the particle executes an ever decreasing spiral around the horizon without ever crossing it. This is reminiscent of the behaviour of a particle approaching the event horizon of a Schwarzschild black hole. Now what happens from the point of view of the infalling particle as it approaches the horizon? From (3.33)$dr/d\tau$ remains finite up to and at the horizon, so the particle crosses the horizon in a finite lapse of its own proper time

τ. Also, as the horizon is approached $d\phi/d\tau \to \infty$, so an infalling observer will see the heavens spinning overhead ever faster and infinitely fast at the horizon. These two apparently contradictory accounts can be reconciled because at the horizon an infalling clock measuring proper time τ is running infinitely more slowly than a clock at infinity measuring time t, as can be seen from equation (3.30).

Problem 40 *If a particle is released from rest at infinity i.e. with $E = 1$ show that it spirals in on a cone of constant half angle.*

3.8.2 Orbits with non-zero angular momentum

So far we have considered particles with zero angular momentum. We can find the angular velocity of a particle with non-zero angular momentum as follows. As before, for the angular speed of the particle seen from infinity we have

$$\Omega = \frac{d\phi}{dt} = \frac{u^3}{u^0} = \frac{g^{33}u_3 + g^{30}u_0}{g^{00}u_0 + g^{03}u_3}.$$

We rewrite this in terms of $\omega = g^{30}/g^{00}$:

$$\Omega = \omega - \frac{g^{30}}{g^{00}} + \frac{g^{33}u_3 + g^{30}u_0}{g^{00}u_0 + g^{03}u_3} = \omega - \frac{[(g^{30})^2 - g^{00}g^{33}]u_3}{(g^{00})^2(u_0 + \omega u_3)},$$

or, finally, using the identities in section 3.3.2

$$\Omega = \omega - \frac{\rho^4 \Delta}{A^2 \sin^2\theta} \frac{u_3}{(u_0 + \omega u_3)}. \tag{3.34}$$

This result applies both to massive and massless particles, with u_3 negative for co-rotation and positive for counter-rotation of the particle with respect to the hole. Note that at the horizon ($\Delta = 0$) all particles, irrespective of their angular momentum, orbit with angular velocity ω.

For the special case of a particle projected in the azimuthal direction in the equatorial plane the 4-velocity of the particle is $u_\mu = (u_0, 0, 0, u_3)$. Expanding the scalar product $g^{\mu\nu}u_\mu u_\mu = 1$ gives

$$g^{00}(u_0)^2 + 2g^{03}u_0 u_3 + g^{33}(u_3)^2 - 1 = 0.$$

Now solve this quadratic for u_0 to get

$$u_0 = -\omega u_3 \pm \left[\left((g^{03})^2 - g^{00}g^{33}\right)u_3^2 + g^{00}\right]^{1/2}/g^{00}.$$

Using the last identity from section 3.3.1 and $L_z = -u_3$ gives

$$u_0 + \omega u_3 = \frac{r^2 \Delta^{1/2}}{A}\left(1 + \frac{A}{r^2 L^2}\right)^{\frac{1}{2}} u_3.$$

Substituting for $u_3/(u_0 + \omega u_3)$ in equation (3.34) gives

$$\Omega = \omega \pm \frac{r^2 \Delta^{\frac{1}{2}}}{A \left(1 + \frac{A}{r^2 L^2}\right)^{\frac{1}{2}}}. \tag{3.35}$$

This is the instantaneous angular velocity of a particle projected in the azimuthal direction having angular momentum per unit mass $L_z = L$. The limiting case of a photon occurs when $L \to \infty$. We shall use this result in section 3.10.2.

3.9   Zero angular momentum observers (ZAMOs)

In Schwarzschild spacetime we used a convenient set of local observers that 'hovered' at fixed coordinate positions. These observers have four-velocity

$$(u^\mu) = (dt/d\tau, 0, 0, 0) = ((g_{00})^{-1/2}, 0, 0, 0)$$

and their world lines are orthogonal to surfaces of constant coordinate time t: if $(dx^\mu) = (0, dr, d\theta, d\phi)$ is a displacement in the surface $t = $ constant then $u_\mu dx^\mu = 0$.

In the Kerr spacetime the world line of a hovering observer having fixed spatial coordinates is not orthogonal to surfaces of constant Boyer-Lindquist time because of the non-zero cross term in the metric: we have

$$u_\mu dx^\mu = g_{\mu\nu} u^\nu dx^\mu = g_{03}(g_{00})^{-1/2} d\phi \neq 0.$$

Therefore, for the hovering observers, events at the same (Boyer-Lindquist) time t are not simultaneous. In order to have a convenient set of observers for which events at the same time t are simultaneous, we define a class of zero angular momentum observers (ZAMOs). These observers have angular speed ω about the spin axis, maintaining their r and θ coordinates constant. So they are not in free fall. They correspond to the hovering observers in the Schwarzschild case. We shall first find their four velocity and then demonstrate that it is orthogonal to the surfaces of constant t.

Consider an observer in a circular orbit with zero angular momentum: $L_z = 0$, hence $u_3 = 0$. Then

$$0 = u_3 = g_{33} u^3 + g_{03} u^0,$$

or

$$u^3 = -\frac{g_{30}}{g_{33}} u^0. \tag{3.36}$$

But from the identities of section 3.3.2 $-g_{30}/g_{33} = g^{30}/g^{00}$, which from problem 37 is just the angular velocity ω of an inertial frame. Thus $u^3 = \omega u^0$ for a ZAMO.

To find u^0 we use $1 = u^\mu u_\mu = u^0 u_0$ (since $u_3 = u_2 = u_1 = 0$) and

$$u_0 = g_{00} u^0 + g_{03} u^3 = (g_{00} - \frac{g_{30}}{g_{33}} g_{03}) u^0.$$

This gives

$$(u^0)^2 = \frac{g_{33}}{g_{00}g_{33} - (g_{03})^2} = -\frac{g_{33}}{\Delta \sin^2 \theta}$$

and hence

$$(u^\mu)_{\text{ZAMO}} = \left(\frac{-g_{33}}{\Delta \sin^2 \theta}\right)^{1/2} (1, 0, 0, \omega). \qquad (3.37)$$

We can now compute the scalar product $dx_\mu (u^\mu)_{\text{ZAMO}}$:

$$dx_\mu (u^\mu)_{\text{ZAMO}} = (g_{03}u^0 + g_{33}u^3)d\phi = 0$$

from (3.36). If follows that the world lines of zero angular momentum observers are orthogonal to the surfaces of constant t.

Problem 41 *For $a = 0$ show that $(u^\mu)_{\text{ZAMO}}$ is the 4-velocity of the Schwarzschild hovering observers.*

3.9.1 Some applications of ZAMOs

We find the local energy per unit mass relative to a ZAMO, γ_{ZAMO}, of a freely falling particle with zero angular momentum at some radius r. The 4-velocity of a zero angular momentum particle has the form $(u_\mu) = (E, u_r, u_\theta, 0)$ and hence

$$\gamma_{\text{ZAMO}} \equiv (u^\mu)_{\text{ZAMO}} u_\mu$$
$$= E \left(\frac{g_{33}}{-\Delta \sin^2 \theta}\right)^{1/2}.$$

As we approach the horizon, $\Delta \to 0$, so $\gamma_{\text{ZAMO}} \to \infty$ i.e. the particle moves towards the horizon with a speed approaching the speed of light. This is just the result we found for the velocity of a freely falling particle with respect to a hovering observer in the Schwarzschild metric.

Problem 42 *Show that the velocity of a ZAMO with respect to a hovering observer in the equatorial plane is*

$$v = \frac{2ma}{r\Delta^{1/2}}$$

in the ϕ-direction. [Hint: $u^\mu_{\text{ZAMO}} u_\mu = \gamma_{\text{ZAMO}} = (1 - v^2)^{-1/2}$, and the motion is entirely in the ϕ-direction.] The speed is unity (i.e. the speed of light) at the surface $r = 2m$, which we shall see later is the static limit surface.

As a second example we shall find the energy per unit mass $E = u_0$ relative to an observer at infinity of a ZAMO for which by definition $u_3 = 0$. The time component of $u^\mu = g^{\mu\nu}u_\nu$ is

$$u^0 = g^{00}u_0 + g^{03}u_3 = g^{00}E.$$

Hence

$$E = \left(\frac{-g_{33}}{\Delta \sin^2 \theta} \right)^{\frac{1}{2}} \left(\frac{\rho^2 \Delta}{A} \right) = \left(\frac{\rho^2 \Delta}{A} \right)^{\frac{1}{2}},$$

where u^0 comes from equation (3.37).

Note that with $a = 0$ and $\theta = \pi/2$, $E = (1 - 2m/r)^{1/2}$, which is the energy of a hovering observer in the Schwarzschild metric. This futher emphasises that ZAMOs are the natural extension of the Schwarzschild hovering observer to the Kerr metric. Notice also that the energy per unit mass of a ZAMO at the horizon is zero. We shall use this result later in section 3.15.4.

3.10   Photon orbits

We now consider the motion of photons in the Kerr geometry. The equations of motion are obtained in a manner analogous to that for massive particles. From the symmetries of the metric $k^\mu p_\mu =$ constant, where p is the momentum 4-vector of a photon and k is a vector pointing along a direction of symmetry. For the Kerr metric directions of symmetry are $(k^\mu) = (1, 0, 0, 0)$ and $(k^\mu) = (0, 0, 0, 1)$ so we have

$$p_0 = E_{ph}$$
$$p_3 = -(L_{ph})_z$$

where E_{ph} and $(L_{ph})_z$ are respectively the conserved energy and the z-component of angular momentum.

3.10.1   The photon effective potential

Multiplying Eq.(3.23) by m_0^2, and specialising to the equatorial plane, the radial equation for a particle of mass m_0 in the equatorial plane has the form

$$m_0^2 r^4 \left(\frac{dr}{d\tau} \right)^2 = r^2 \left(r^2 + a^2 + \frac{2ma^2}{r} \right) m_0^2 E^2 - 4marm_0^2 E L_z$$
$$- r^2 \left(1 - \frac{2m}{r} \right) m_0^2 L_z^2 - m_0^2 \Delta r^2.$$

Taking the limit $m_0 \to 0$ and $d\tau \to 0$ with $d\lambda = d\tau/m_0$ finite, we obtain the radial equation for light rays (i.e. the equation of motion for photons) in the equatorial plane of the form,

$$r^4 \left(\frac{dr}{d\lambda} \right)^2 = r^2 \left(r^2 + a^2 + \frac{2ma^2}{r} \right) E_{ph}^2 - 4mar E_{ph} L_{ph} - r^2 \left(1 - \frac{2m}{r} \right) L_{ph}^2. \quad (3.38)$$

where $E_{ph} = m_0 E$ and $L_{ph} = m_0 L_z$ are the conserved energy and angular momentum for light.

Following the derivation of the effective potential for a massive particle is section 3.7.6, the effective potential for a photon in

$$V_\pm = \frac{2mra \pm r^2\Delta^{1/2}}{(r^2+a^2)^2 - a^2\Delta} L_z.$$

Eq. (3.38) can then be written as

$$r^4\left(\frac{dr}{d\lambda}\right)^2 = A(E - V_+)(E - V_-). \tag{3.39}$$

From this it is clear that if $\Delta < 0$ then $V_\pm$ becomes complex and $dr/d\lambda$ cannot be zero. Therefore, once a photon has moved inside $\Delta = 0$, (r decreasing, $dr/d\lambda < 0$), it cannot turn round and escape to larger r. Thus we have shown that $\Delta = 0$ is the event horizon for motion in the equatorial plane of the Kerr solution.

3.10.2   Azimuthal motion

A light ray projected in the azimuthal direction in the equatorial plane has an instantaneous angular velocity given by (3.35) in the limit as $L \to \infty$

$$\Omega = \omega \pm \frac{r^2\Delta^{1/2}}{A}. \tag{3.40}$$

Note that at the horizon $\Delta = 0$ so all particles rotate with the hole at a rate $\Omega = \omega_H$ (given by Eq. (3.29)), irrespective of their angular momentum.

We also see the possibility that counter-rotating photons at some radius can have zero angular velocity as seen from infinity. This means that the frame-dragging there is so strong that even photons cannot 'move backwards' (i.e. against the direction of spin of the hole). In the equatorial plane this occurs at a radius given by $A\omega = r^2\Delta^{1/2}$ or

$$A^2\omega^2 = 4m^2a^2r^2 = r^4\Delta = r^4(r^2 + a^2 - 2mr),$$

which is satisfied by $r = 2m$ (and hence $\Delta = a^2$). The surface in which counter-rotating photons appear stationary from infinity is called the static limit surface. (See section 3.11.) We have therefore shown that the intersection of the static limit surface with the equatorial plane occurs at $r = 2m$.

3.10.3   Photon capture cross-section

Now we can look at the effect of rotation on the capture of photons by a black hole. This can done by considering circular photon orbits, which are unstable and lie on the borderline between capture by the hole and escape to infinity. For simplicity we confine our attention to motion in the equatorial plane. As in section 2.3.9 the conditions for circular orbits are obtained from the radial equation of motion. From equation (3.22) for photons moving in the equatorial plane we get as

$$\left(\frac{dr}{d\lambda}\right)^2 = r^{-4}\{\left[E_{ph}(r^2 + a^2) - aL_{ph}\right]^2 - \Delta(L_{ph} - aE_{ph})^2\}$$

$$= E_{ph}^2 + \frac{2m}{r^3}(L_{ph} - aE_{ph})^2 - \frac{1}{r^2}(L_{ph}^2 - a^2E_{ph}^2)$$

$$= R(r).$$

The conditions for a circular orbit are, as before (section 2.3.9), $dr/d\lambda = 0$ and $dR/dr = 0$. Applying these conditions for co-rotating photons gives

$$E_{ph}^2 + \frac{2m}{r_{ph}^3}(L_{ph} - aE_{ph})^2 - \frac{1}{r_{ph}^2}(L_{ph}^2 - a^2E_{ph}^2) = 0,$$

and

$$-\frac{6m}{r_{ph}^4}(L_{ph} - aE_{ph})^2 + \frac{2}{r_{ph}^3}(L_{ph}^2 - a^2E_{ph}^2) = 0.$$

We can deal with counter-rotating photons by changing the sign of a. The second of these gives directly

$$r_{ph} = 3m\frac{b - a}{b + a},$$

where $b = L/E$ is the impact parameter below which capture will occur. For $a = 0$ we regain the Schwarzschild result $r_{ph} = 3m$; for a co-rotating hole we have $a > 0$ and $r_{ph} < 3m$, with the opposite inequalities for a counter-rotating hole.

Solving for the impact parameter in terms of the radius of the circular photon orbit

$$b = -a\frac{r_{ph} + 3m}{r_{ph} - 3m},$$

so the extreme values are $b = 2m$ (for $a = m$, $r_{ph} = m$ - see below, section 3.13.6) and $b = 7m$ (for $a = -m$, $r_{ph} = 4m$). These straddle the Schwarzschild value $b = 3\sqrt{3}m$. Thus, contra-rotating photons are more likely to be captured than co-rotating photons. We shall refer to this result in section 3.16.1 below.

The capture cross-section for photons incident perpendicular to the spin axis is $\sigma_{\text{perp}} \approx 24.3\pi m^2$ and for photons moving in a direction parallel to the spin axis is $\sigma_{\parallel} \approx 23.3\pi m^2$. (See Frolov and Novikov, p78, for further details.) Thus the capture cross-section for a rotating black hole is less than $\sigma = 27\pi m^2$ for a non-rotating one (section 2.7.1).

3.11 The static limit surface

We shall show that the static limit surface is defined by $g_{00} = 0$, from which we shall be able to find this surface in general (not just in the equatorial plane). Consider a stationary particle, r = constant, θ =constant, ϕ = constant. We have

$$1 = g_{00}\left(\frac{dt}{d\tau}\right)^2.$$

For $g_{00} \leq 0$ this condition cannot be satisfied, so a massive particle cannot be stationary within the surface $g_{00} = 0$. For photons we can satisfy the corresponding condition if $g_{00} = 0$, which is in agreement with what we found above, that photons can be stationary at the static limit. Solving the condition $g_{00} = 0$ for r gives us the radius of the static limit surface:

$$r_{st} = m + (m^2 - a^2\cos^2\theta)^{1/2},$$

where we take the positive square root to obtain the outer surface, and also for agreement with $r_{st} = 2m$ in the equatorial plane.

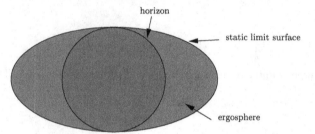

Figure 1   Ergosphere of a Kerr black hole.

Note that the static limit surface lies outside the horizon $\Delta = 0$, except at the poles where the two surfaces coincide (see Fig. 1). We show next that the static limit surface in the Kerr solution is not a horizon. This is in contrast to the Schwarzschild geometry, where the event horizon and the static limit surface coincide. Consider a particle in a circular orbit (not necessarily a geodesic) on θ = constant. We have

$$\begin{aligned} 1 &= g_{00}(u^0)^2 + 2g_{03}u^0u^3 + g_{33}(u^3)^2 \\ &= (u^0)^2(g_{00} + 2\Omega g_{03} + \Omega^2 g_{33}), \end{aligned}$$

where as usual $\Omega = d\phi/dt = u^3/u^0$. Thus, for the motion to be possible, we require $g_{00} + 2\Omega g_{03} + \Omega^2 g_{33} > 0$. Inside the static limit we have $g_{00} \leq 0$ and (as usual outside the horizon) $g_{33} < 0$, so the inequality can be satisfied only if $2\Omega g_{03} > -\Omega^2 g_{33}$, or if

$$4mar\sin^2\theta > \Omega[(r^2 + a^2)^2 - a^2\Delta\sin^2\theta].$$

As an example, this inequality is satisfied for a zero angular momentum particle, since in this case $\Omega = \omega = 2mar\sin^2\theta/[(r^2+a^2)^2 - a^2\Delta\sin^2\theta]$. Thus inside the static limit there are possible circular orbits. We can perturb a circular orbit to give the particle a small velocity in the outward radial direction. Thus a particle can escape from within the static limit surface, so it is not an event horizon.

Problem 43 *Show that the purely radial motion of light inside the static limit is not possible.*

Problem 44 *Show that a stationary particle (i.e. a particle with fixed coordinates) in the Kerr metric has angular momentum per unit mass*

$$L_z = -\frac{2mar\sin^2\theta}{\rho(\Delta - a^2\sin^2\theta)^{1/2}}$$

in the opposite sense to the rotation of the hole. Hence show that it is impossible for a body to remain stationary inside the static limit surface.

3.12 The infinite redshift surface

The redshift of radiation emitted by a body at rest at radius r and observed at infinity can be obtained as follows. The metric relates the local time τ of the stationary frame at r to the time t for a stationary observer at infinity. The redshift is given in terms of the wavelengths of the radiation by

$$1 + z = \frac{\lambda_\infty}{\lambda_r}$$

so, since the wavelength is proportional to the period of the wave, the redshift also relates the two times:

$$1 + z = \frac{dt}{d\tau} = (g_{00})^{-1/2}.$$

Thus the static limit surface in the Kerr metric, $g_{00} = 0$, is also an infinite redshift surface, just as in Schwarzschild. The coincidence of these two surfaces is peculiar to these geometries and is not true for general black holes (for example, in the presence of external matter).

3.13 Circular orbits in the equatorial plane

We now consider circular orbits in the equatorial plane. As for the Schwarzschild case we shall find that there are stable bound orbits down to a limiting value of radius r inside which the circular orbits are unstable. At still smaller radii the unstable orbits become unbound and culminate in a circular photon orbit. The location of the transition between these different types of orbits depends on the angular momentum of the black hole a and the sense of rotation of the orbit (prograde or retrograde) with respect to the hole.

The radial motion of a particle is governed by (3.23)

$$\left(r^2\frac{dr}{d\tau}\right)^2 = R(E, L, r) \tag{3.41}$$

with $Q = 0$ and $L = L_z$. For a circular orbit we require that the radius is constant, hence that $dr/d\tau = 0$, and also that $d^2r/d\tau^2 = 0$. Thus, for a circular orbit as we saw in section 2.3.9 $R = 0$ and $dR/dr = 0$. After some rearrangment of equation (3.23) the condition $R = 0$ gives

$$R = \left(r^2 + a^2 + \frac{2ma^2}{r}\right)(E^2 - 1) - \frac{4maEL}{r} - \left(1 - \frac{2m}{r}\right)L^2 + 2m\left(r + \frac{a^2}{r}\right) = 0$$
(3.42)

and the condition $dR/dr = 0$ becomes

$$\frac{dR}{dr} = 2\left(r - \frac{ma^2}{r^2}\right)(E^2 - 1) + \frac{4maEL}{r^2} - \frac{2mL^2}{r^2} + 2m\left(1 - \frac{a^2}{r^2}\right) = 0. \quad (3.43)$$

The idea is to solve these equations for E and L (Lynden-Bell 1978). The first step is to eliminate the cross term from (3.42) and (3.43) giving

$$R + r\frac{dR}{dr} = (E^2 - 1)(3r^2 + a^2) + 4mr - L^2 = 0, \quad (3.44)$$

which is an explicit relation between E and L. Eliminating L from (3.43) using (3.44) gives a quadratic for $E^2 - 1$. After some considerable algebra we obtain

$$E = \frac{1 - \frac{2m}{r} \pm \frac{a}{m}\left(\frac{m}{r}\right)^{3/2}}{\left[1 - \frac{3m}{r} \pm 2\frac{a}{m}\left(\frac{m}{r}\right)^{3/2}\right]^{1/2}}. \quad (3.45)$$

Substituting (3.45) into (3.44) gives

$$L = \pm\frac{m}{\left(\frac{m}{r}\right)^{1/2}}\frac{1 + \left(\frac{a}{m}\right)^2\left(\frac{m}{r}\right)^2 \mp 2\frac{a}{m}\left(\frac{m}{r}\right)^{3/2}}{\left[1 - \frac{3m}{r} \pm 2\frac{a}{m}\left(\frac{m}{r}\right)^{3/2}\right]^{1/2}}. \quad (3.46)$$

The upper signs correspond to prograde orbits (in the same direction as the spin of the hole) and the lower signs to retrograde orbits.

3.13.1 Innermost (marginally) stable circular orbit

The stability of a circular orbit can be shown to depend on the sign of the second derivative of the effective potential, as one might expect. To simplify the algebra however, it is easier to proceed directly as follows. Writing L for L_z, the radial equation in the equatorial plane again has the form (3.41):

$$r^4\left(\frac{dr}{d\tau}\right)^2 = R(r, L, E),$$

where, from (3.23) with $Q = 0$, the function R is

$$R = \{[E(r^2 + a^2) - aL]^2 - \Delta[r^2 + (L - aE)^2]\}/r^2.$$

We can derive the condition for stability in the standard way. Assume that the particle is in the orbit $r = r_0$ and make a small perturbation to the orbit $r = r_0 + \varepsilon$. Then

$$r_0^4 \left(\frac{d\varepsilon}{d\tau}\right)^2 = R(r_0) + \varepsilon R'(r_0) + \frac{1}{2}\varepsilon^2 R''(r_0) + \dots,$$

where the prime denotes differentiation with respect to r. The orbit $r = r_0$ is a solution of $R(r_0) = 0$ and $R'(r_0) = 0$, so ε grows exponentially

$$d\varepsilon/d\tau \simeq [R''(r_0)/2r_0^4]^{1/2}\varepsilon \propto +\varepsilon$$

unless $R''(r_0) \leq 0$. For the marginally stable case we have $R''(r_0) = 0$. This allows us to determine the energy, angular momentum and radius of the innermost stable circular orbit (from the three conditions on R).

To simplify the algebra the trick is to note that equation (3.44) has the form $R + rR' = 0$. Thus we can differentiate again to obtain

$$2R' + rR'' = 6r(E^2 - 1) + 4m \qquad (3.47)$$

an expression that must vanish at the radius $r_0 = r_{ms}$ of a marginally stable orbit, where both $R' = 0$ and $R'' = 0$. The resulting equation (3.47) is readily solved for E^2:

$$E^2 = 1 - \frac{2}{3}\frac{m}{r_{ms}}. \qquad (3.48)$$

But we also have the condition (3.45) for $r(E)$ on a circular orbit. Eliminating E from (3.45) using (3.48) gives

$$r_{ms}^2 - 6mr_{ms} \pm 8am^{1/2}r_{ms}^{1/2} - 3a^2 = 0. \qquad (3.49)$$

This is a quartic in $r_{ms}^{1/2}$, which can be solved (Bardeen 1972), although the resulting algebraic expression is not very illuminating. Instead consider two special cases. First, we can confirm that we recover the Schwarzschild result $r_{ms} = 6m$ on putting $a = 0$. Second, for the extreme Kerr solution with $a = m$, the quartic for the innermost (marginally) stable orbit at $r = r_{ms}$ reduces to

$$\left(\frac{r_{ms}}{m}\right)^2 - 6\left(\frac{r_{ms}}{m}\right) \pm 8\left(\frac{r_{ms}}{m}\right)^{1/2} - 3 = 0.$$

With the upper sign this is satisfied by $r_{ms} = m$; this is the marginally stable orbit for a co-rotating particle. With the lower sign we see that $r_{ms} = 9m$, which is the solution for a counter-rotating particle.

Problem 45 *It is difficult to solve Eq. (3.49) for r_{ms} but it can be solved readily as an equation for a, the angular momentum of the hole, as a function of r_{ms}. Hence verify that for retrograde orbits $r_{ms} = 7.5m$ corresponds to $a = 0.48m$ and for prograde orbits $r = 4.12m$ corresponds to $a = 0.53m$.*

3.13.2 Period of a circular orbit

The coordinate period of a particle in orbit in the equatorial plane of the Kerr metric is important astrophysically because this is a timescale on which periodic variability in the emission of radiation from material in orbit about the hole, for example in an accretion disc, would be detected by a distant observer. We can calculate this period as follows. We have

$$\frac{d\phi}{dt} = \frac{d\phi}{d\tau}\frac{d\tau}{dt},$$

where the quantities on the right are independent of time t and angle ϕ. Thus, integrating over a period T, $0 \le \phi \le 2\pi$,

$$2\pi = \left(\frac{d\phi}{d\tau}\frac{d\tau}{dt}\right) T$$

or

$$T = 2\pi \frac{dt}{d\tau} \Big/ \frac{d\phi}{d\tau}.$$

Using equations (3.14) and (3.18) with E and L given by (3.45) and (3.46) we get, after some algebra,

$$T = 2\pi m \left(\frac{r}{m}\right)^{3/2} \frac{\left[1 + \frac{a^2}{r^2} - \frac{2m}{r} \pm \left(\frac{a}{m}\right)\left(\frac{m}{r}\right)^{3/2} \pm \left(\frac{a}{m}\right)^3 \left(\frac{m}{r}\right)^{7/2} \mp 2\left(\frac{a}{m}\right)\left(\frac{m}{r}\right)^{5/2}\right]}{\left[1 + \frac{a^2}{r^2} - \frac{2m}{r}\right]}.$$

$$(3.50)$$

which, on dividing the numerator by the denominator, simplifies to

$$T = 2\pi \left(\frac{r^{\frac{3}{2}}}{m^{\frac{1}{2}}} \pm a\right).$$

In the Schwarzschild case, $a = 0$, we get

$$T = 2\pi m \left(\frac{r}{m}\right)^{3/2}$$

and the last stable orbit is at $r_{ms} = 6m$, so the minimum period is $T \approx 29.4\pi m$. For the black hole at the centre of the Galaxy, with a mass $4.0 \times 10^6 M_\odot$, this period is about 30 minutes. At the opposite extreme, if the black hole at the Galactic centre were an extreme Kerr hole with material in prograde orbits then $a = m$ and $T = 4\pi m$, giving a period of 3.8 mins. For extreme retrograde orbits, then $a = m$, $r_{ms} = 9m$ and $T = 52\pi m$, giving a period of about 48 minutes. Note that most of the difference arises from the change to r_{ms} from $6m$ to $9m$ rather than the explicit dependence of T on a in Eq. (3.50).

Problem 46 *Find the period of a particle in the last stable prograde orbit about the black hole at the centre of the Galaxy assuming $a = 0.53$ and $r = 4.12$.*

3.13.3   Energy of the innermost stable orbit

Given a value for a we can solve the quartic equation (3.49) to get the radius r_{ms} of the innermost stable orbit. Substituting the value of r_{ms} into equation (3.48) gives the energy per unit mass of a particle in this orbit. For the limiting case of co-rotation around an extreme Kerr black hole we have $r_{ms} = m$ and thus,

$$E = \left(1 - \frac{2}{3}\right)^{1/2} = \frac{1}{\sqrt{3}}.$$

The binding energy of a particle in an orbit is the difference between the orbital energy and the energy of the particle at infinity. So the binding energy per unit mass is $1 - E$, or about 42 per cent of the energy of the particle at rest at infinity. The corresponding results for a counter-rotating particle are $r_{ms} = 9m$ for the radius of the innermost stable orbit and $E = (25/27)^{1/2}$, and a binding energy of 3.8 per cent. For comparison, recall that for a non-rotating black hole the corresponding binding energy is 5.7 per cent of the rest mass energy.

    In fact we shall see in sections (3.16.1) and (4.4.2) that real black holes cannot be spun up to the extreme Kerr limit. A more realistic upper limit to a is given by $a = 0.998m$. Inserting this value into equation (3.48) gives $r_{ms} = 1.237m$, which yields a maximum binding energy for a co-rotating particle of about 32 per cent. This is the maximum energy that can be realistically extracted from matter in an accretion disc as it spirals into a black hole.

Problem 47 *Verify, using equation (3.49), that $r_{ms} = 1.237m$ for $a = 0.998m$. Hence, using Fig. 2, sketch the relationship between binding energy in the innermost stable orbit and a.*

3.13.4   Angular momentum of the innermost stable orbit

Substituting for E^2 from (3.48) in (3.44) we can obtain the angular momentum on the innermost (marginally) stable circular orbit. Using (3.49) we can eliminate a to get

$$L = \pm \frac{2m}{3\sqrt{3}} \left\{ 1 + 2 \left[3 \left(\frac{r_{ms}}{m} \right) - 2 \right]^{1/2} \right\}. \tag{3.51}$$

Problem 48 *Derive equation (3.51).*

Problem 49 *Find the angular momentum carried into an extreme Kerr black hole from the marginally stable orbit by (a) a co-rotating particle and (b) a counter-rotating particle.*

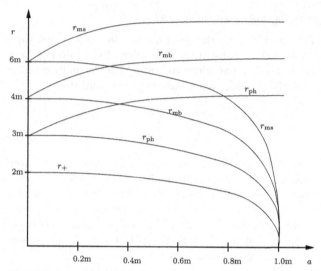

Figure 2   Radii of circular orbits in the equatorial plane as a function of a/m. The upper curves refer to retrograde and the lower to prograde orbits. (From Bardeen *et al.* 1972.)

3.13.5   Marginally bound orbits

Bound orbits correspond to $E \leq 1$. We can calculate the radius of a marginally bound orbit by putting $E = 1$ in (3.45). Solving the resulting quartic equation gives

$$r_{mb} = 2m \mp a + 2m^{1/2}(m \mp a)^{1/2}$$

with the upper signs corresponding to co-rotation (Bardeen, 1972). As an example, if $a = m$ for an extreme Kerr black hole, then $r_{mb} = m$ or $(3 + 2\sqrt{2})m$.

3.13.6   Unbound orbits

Circular orbits exist where the square root in the denominator of equation (3.45) is real, hence where

$$\left(\frac{r}{m}\right)^{3/2} - 3\left(\frac{r}{m}\right)^{1/2} \pm 2\left(\frac{a}{m}\right) \geq 0. \tag{3.52}$$

Solving this cubic in the limiting case of equality gives

$$r = r_{ph} = 2m\left\{1 + \cos\left[\frac{2}{3}\cos^{-1}\left(\mp\frac{a}{m}\right)\right]\right\}.$$

In this case the energy per unit mass $E \to \infty$, so the orbit must (if anything) correspond to a zero mass particle i.e. a photon orbit. For $a = 0$ we recover the Schwarschild result, $r_{ph} = 3m$, for the innermost photon orbit. For the extreme Kerr solution we get $r_{ph} = m$, or $4m$ (co-rotation or counter-rotation respectively). Just as in the Schwarzschild case we see that there is a region in which circular orbits can be unbound!

Note that although in the extreme Kerr solution $r_{ph} = r_{mb} = r_{ms} = m$ these orbits are distinct and separated from each other by finite proper distances. The explanation lies with the failure of the Boyer-Lindquist r coordinate to represent the geometry in this region i.e. the coordinate system becomes singular at this radius (Bardeen et al., 1972). See Figs. 2 and 3.

Figure 3 Embedding diagram for the Kerr solution. As $a/m \to 1$, the 'throat' approaches a cylinder: the radii of the various labelled surfaces (shown as curves in this representation) become equal in magnitude, but the surfaces are nevertheless distinct.

Problem 50 *Show that the impact parameter $b = -a\frac{r+3m}{r-3m} = 3\sqrt{3}m$ for a Schwarzschild black hole (section 3.10.3). Hint: use equation (3.52) in the case of equality and solve for a.*

3.14 Polar orbits

We get another view of the effects of frame dragging by a rotating black hole if we look at the precession of non-equatorial orbits. To state this precisely, we define the line of the ascending node as the line from the origin in the equatorial plane through the point at which the orbit crosses the plane in going from negative to positive latitudes. For orbits of decreasing radii we find that the line of the ascending node is increasingly dragged in the sense of the spin of the black hole. To see how this comes about, consider the simple case of a circular orbit that passes through the poles. The angle along the orbit is then θ and the precession is measured by ϕ.

For an orbit through the poles $L_z = 0$, so the radial equation Eq. (3.22) becomes

$$\rho^4 \left(\frac{dr}{d\tau}\right)^2 = E^2(r^2 + a^2)^2 - \Delta\left[r^2 + a^2 E^2 + Q\right] \equiv V_r.$$

The condition for a circular orbit is, as usual, $V_r = 0$ and

$$\frac{dV_r}{dr} = 4r\left(r^2 + a^2\right) - 2r\Delta - 2\left(r - m\right)\left[r^2 + a^2E^2 + Q\right] = 0.$$

Eliminating the expression in square brackets between these two equations gives

$$E^2 = \frac{r\Delta^2}{\left(r^2 + a^2\right)\left(r^3 - 3mr^2 + a^2r + a^2m\right)} \tag{3.53}$$

for the energy of a particle in a circular polar orbit.

Only orbits having $E^2 > 0$ can exist. The numerator of (3.53) is positive for all values of r down to the horizon but the denominator will fall to zero for some value of r before the horizon is reached at which point $E^2 \to \infty$: this is the circular photon orbit. We can find the r coordinate of the photon orbit as a function of a by solving the cubic equation

$$r^3 - 3mr^2 + a^2r + a^2m = 0.$$

For the limiting case of an extreme Kerr black hole it has the value $r_{ph} = (1 + \sqrt{2})m$.

Problem 51 *For large r ($\gg m$ and $\gg a$) show that we recover the Schwarzschild result*

$$E^2 = \frac{(1 - 2m/r)^2}{(1 - 3m/r)}. \tag{3.54}$$

See section 2.3.10.

Consider now the motion of a particle in the ϕ direction. For $L_z = 0$ Eq. (3.16) gives

$$\rho^2\frac{d\phi}{d\tau} = \frac{2marE}{\Delta}.$$

Note that there is no precession if $a = 0$, so the precession is entirely due to the spin of the hole. To estimate the precession rate we use the approximation of large r (or small a) for ρ, E, and Δ. This gives us

$$\omega_p = \frac{d\phi}{d\tau} \approx \frac{d\phi}{dt} \approx \frac{2ma}{r^3}. \tag{3.55}$$

for the precession frequency. The dragging of the orbital plane of a non-equatorial orbit is known as the Lens-Thirring effect.

The mental picture that the term frame dragging conjures up is that of a whirlpool dragging everything around in the direction of its sense of rotation. This picture should be treated with caution as it can sometimes mislead us. As we have just seen the picture works for the case of polar orbits which are dragged around in the sense of the rotation of the hole. It also works for the case of a zero angular momentum particle in free fall in the equatorial plane (section 3.8). On the other hand the whirlpool picture does not help us when we consider particles in circular orbits in the equatorial plane. We saw in section 3.13.1 that for circular orbits having $r > 9m$ particles can orbit in either sense about the black hole. But surprisingly, for an orbit at a given radial coordinate, it is the particle orbiting in the opposite sense to the rotation of the hole that goes fastest. This can be seen from Eq. (3.50) and is the contrary of what the whirlpool analogy might lead us to expect.

Problem 52 *For an extreme Kerr black hole ($a = m$) plot E from equation (3.53) as a function of r to find numerically the minimum value of E and hence the maximum binding energy.*

Problem 53 *Eq. (3.55) is given in geometrical units so the time τ is in metres, and ω_p is in m^{-1}. Show that ω_p is given in SI units by*

$$\omega_p(in\ seconds^{-1}) = \frac{2G}{c^2}\frac{J}{r^3},$$

where J is the angular momentum of the black hole ($= aMc$ for a body of mass M). Show that for a given r/m (i.e. for a given distance from the hole measured in terms of Schwarzschild radii) the precession is stronger for lower mass black holes.

Problem 54 *A satellite is in a polar orbit around the Earth at an altitude of 12 000 km. Show that the plane of the orbit will be dragged round by the rotation of the Earth at a rate of about 3×10^{-2} arcsec per year. (The moment of inertia of the Earth is $0.33 M_\oplus R_\oplus^2$.)*

Problem 55 *Show that at large distances from a slowly rotating body a particle in the equatorial plane having zero angular momentum is dragged round at the same angular frequency as that at which a polar orbit of the same radius precesses.*

Problem 56 *Using Eq. (3.53) and the condition $V_r = 0$ show that for a circular polar orbit*

$$Q = \frac{mr^2\left(r^4 + 2a^2r^2 - 4ma^2r + a^4\right)}{\left(r^2 + a^2\right)\left(r^3 - 3mr^2 + a^2r + a^2m\right)}$$

and hence, for $a = 0$ (the Schwarzschild case, valid for small a or large r)

$$Q = \frac{mr}{(1 - 3m/r)}.$$

3.14.1 Orbital period

We can get the period of a circular polar orbit, and hence the precession per orbit, from the θ−equation (3.18), which, for $L_z = 0$ is

$$\rho^4 \left(\frac{d\theta}{d\tau}\right)^2 = Q - a^2(1 - E^2)\cos^2\theta.$$

For large r we get

$$\frac{d\theta}{d\tau} \approx \frac{Q^{1/2}}{r^2}.$$

Using the values of Q from the previous section, again for large r, we obtain

$$\frac{d\theta}{d\tau} \approx \frac{m^{1/2}}{r^{3/2}},$$

giving an orbital period of $2\pi r^{3/2}/m^{1/2}$. So, using equation (3.55), in one orbital period the line of nodes advances by

$$\Delta\phi \approx 2\pi \frac{2ma}{r^3} \frac{r^{3/2}}{m^{1/2}} = \frac{4\pi m^{1/2}a}{r^{3/2}}.$$

This Lens-Thirring effect has been measured to an accuracy of about 20 per cent using the two laser ranging satellites LAGEOS and LAGEOS II in orbit about the Earth. Short laser pulses aimed at the satellites are reflected back by corner cubes giving the distance to an accuracy of a few millimetres. (See Cuifolini *et al*, 2007.)

3.15   The ergosphere

Despite the fact that a black hole is shielded by an event horizon it is nevertheless possible to extract its rotational energy. This implies that the rotational energy is located outside the event horizon. Energy extraction is achieved by exploiting the properties of the region between the event horizon and the static limit surface of a Kerr black hole. This region has consequently been named the ergosphere (from the Greek for energy). This unexpected property of spinning black holes relies on the existence of negative energy orbits for particles inside the ergosphere. In this section we demonstrate the existence of these negative energy states. We then describe a method, the Penrose process, that in principle allows the extraction of energy.

3.15.1 Negative energy orbits

First some comments on energy. The specific energy of a particle having four-velocity u^μ as measured locally by an observer with four-velocity u^μ_{obs} is given by $u_\mu u^\mu_{obs}$. This can be a function of position but is always positive. The energy of this particle measured from infinity is $E = u_\mu k^\mu$, where $k^\mu = (1, 0, 0, 0)$ in Boyer-Lindquist coordinates. This energy is a constant for a free particle, independent of position, but need not be positive. Recall that a negative energy $E < 0$, simply means that it takes more than the rest mass energy of the particle to remove it to infinity.

Returning to the equation for radial motion (3.22), confining ourselves to motion in the equatorial plane, we can solve this quadratic for E even in the case that $dr/d\tau \neq 0$. The result is very similar to our expression for V_{eff},

$$E = \frac{2amLr \pm \sqrt{4a^2m^2L^2r^2 - [r^4 + a^2(r^2 + 2mr)][L^2a^2 - (r^2 + L^2)\Delta - \dot{r}^2r^4]}}{r^4 + a^2(r^2 + 2mr)},$$
(3.56)

where we have written $\dot{r}$ for $dr/d\tau$ and L for L_z. From equation (3.15) we see that to keep $dt/d\tau > 0$ (so that the velocity four-vector points to the future) we must have

$$E > \frac{2marL}{r^4 + a^2(r^2 + 2mr)}.$$

So we take the positive square root in (3.56). If $L < 0$ then we have the potential for E to take negative values, provided the contribution from the square root can be kept small enough. The terms under the square root can be rearranged to collect those involving L^2:

$$4a^2m^2L^2r^2 - [r^4 + a^2(r^2 + 2mr)][L^2a^2 - (r^2 + L^2)\Delta - \dot{r}^2r^4]$$
$$= r^4\Delta L^2 + [r^4 + a^2(r^2 + 2mr)](r^2\Delta + \dot{r}^2r^4).$$
(3.57)

The terms that do not involve L^2 can be made small by restricting the component of motion in the radial direction, $\dot{r}$, and by taking $r^2 \ll L^2$. Since L is the angular momentum per unit mass this last condition can always be satisfied by taking the mass of the particle small enough. So consider orbits for which these terms can be neglected. Now, at the static limit surface in the equatorial plane we have $r = r_{st} = 2m$ and $\Delta = a^2$, so inside r_{st} the term $r^4\Delta L^2$ always contributes less than $(2m)^2r^2a^2L^2$, and hence its square root always contributes less than $+2am\,|L|\,r$ to E. Thus,

$$E < (2amLr + 2am\,|L|\,r)/[r^4 + a^2(r^2 + 2mr)]$$

and if $L < 0$ then $E < 0$. So there are orbits inside r_{st} with $L < 0$ for which also $E < 0$, or, in words, within the static limit surface there are retrograde orbits which have negative energy. As we said, this means that the energy required to remove a particle in such an orbit to infinity is greater than its rest mass.

Problem 57 *By finding $k_\mu k^\mu$ show that the 4-vector $(k^\mu) = (1,0,0,0)$ is spacelike inside the static limit surface.*

Using a standard result in special relativity that the scalar product of a space-like vector with a timelike vector can be negative we deduce from problem (57) that the energy $E = p_\mu k^\mu$ of a particle with 4-velocity u^μ (necessarily timelike) can be negative. The following problem addresses the direct proof of this result for a photon by explicit calculation.

Problem 58 *For a photon in an azimuthal direction in the equatorial plane use Eq. (3.40) for Ω, the definition of the angular velocity of zero angular momentum orbits, $\omega = -g_{03}/g_{33}$, and the equation of motion for $dt/d\lambda = p^0$ to show that*

$$E_{ph} = \frac{2amL_{ph}r + |L_{ph}|\,r^2\Delta^{1/2}}{A}.$$

Demonstrate that this is consistent with (3.56) with $\dot{r} = 0$ and with $m_0 = 0$ for a photon. Deduce that for a counter-rotating particle in the ergosphere (where $r^2\Delta^{1/2} < 2amr$) E_{ph} can be negative if L_{ph} is negative.

Note that although the conserved energy per unit mass relative to an observer at infinity E can be negative, the local energy per unit mass γ is always positive.

3.15.2   Energy and angular momentum

We shall show that the angular momentum per unit mass $L_n = |L|$ and energy per unit mass $E_n = |E|$ of a negative energy particle satisfies the relation $E_n \le \omega L_n$, a result we shall need in chapter 4. Starting from the relation

$$g^{\mu\nu}u_\mu u_\nu = 1,$$

we solve the quadratic in u_0 to get

$$u_0 = -\frac{g^{03}}{g^{00}}u_3 \pm \frac{[(g^{03})^2 u_3^2 - g^{00}(g^{11}(u_1)^2 + g^{22}(u_2)^2 + g^{33}(u_3)^2 - 1)]^{1/2}}{g^{00}},$$

which becomes

$$u_0 = -\omega u_3 \pm \frac{\Delta^{1/2}}{A}\left[\frac{\rho^4}{\sin^2\theta}(u_3)^2 + A\Delta(u_1)^2 + A(u_2)^2 + \rho^2 A\right]^{1/2}. \tag{3.58}$$

On the horizon we have $\Delta = 0$ so

$$u_0 = -\omega_+ u_3.$$

As $u_0 = E$ and $u_3 = -L$ this gives

$$E = \omega_+ L$$

So on the horizon positive L gives positive E, $L = 0$ gives $E = 0$ and negative L gives negative E. Away from the horizon we must take the positive square root in Eq. (3.58) to ensure that u_0 is positive for zero angular momentum, $u_3 = 0$. Then, in all cases, $u_0 + \omega u_3 \ge 0$ or

$$E - \omega L \ge 0.$$

For negative angular momentum particles, with $L = -L_n$ and $E = -E_n$ both negative, this reads $E_n \le \omega L_n$.

3.15.3   The Penrose process

Consider a particle of energy per unit mass E that decays into two particles somewhere in the ergosphere where negative energy orbits are possible. Assume further that one of the decay products has positive angular momentum and positive energy E_p, while the other product has both negative angular momentum and energy E_n. So $E = E_p + E_n$ and $E_p > E$. Let the negative energy particle be captured by the hole, thus reducing its mass and angular momentum. The positive energy particle escapes to infinity with more than the initial energy E. As seen by the outside world therefore, this process extracts energy (and angular momentum) from the hole.

3.15.4   Realising the Penrose process

Despite its simplicity, the Penrose process is unlikely to provide a realistic mechanism for the extraction of energy from a black hole. To explain this statement consider the following. The energy per unit mass of a particle in the last stable circular orbit of an extreme Kerr black hole $(E = 1/\sqrt{3})$ is the lowest achievable energy for an orbiting particle entering the ergosphere from the outside. To realise the Penrose process this particle must decay into two fragments, one of which must be projected against the rotation of the hole into a negative energy orbit. The question that arises is, what is the size of the velocity boost required to do this? To answer this, we need to find the velocity with respect to a local ZAMO of a particle in the last stable circular orbit.

According to Eq. (3.37) the 4-velocity of a ZAMO in the equatorial plane is

$$(u^\mu_{ZAMO}) = \left(\left(-\frac{g_{33}}{\Delta} \right)^{1/2}, 0, 0, \omega \left(-\frac{g_{03}}{\Delta} \right)^{1/2} \right). \tag{3.59}$$

The 4-velocity of a particle in a circular orbit has the form

$$(u_\mu) = (u_0, 0, 0, u_3) = (E, 0, 0, \pm L).$$

Therefore the energy per unit mass in the ZAMO frame is

$$u^\mu_{ZAMO} u_\mu = \left(\frac{-g_{33}}{\Delta} \right)^{1/2} (u_0 + \omega u_3),$$
$$= \gamma,$$

where γ is the Lorentz gamma-factor of the particle with respect to a ZAMO.

We now write the expression for the energy in the ZAMO frame in terms of u_3 by using equation (3.58). For motion in the equatorial plane with r constant this gives

$$u_0 + \omega u_3 = \left(r^2 u_3^2 + A \right)^{1/2} \frac{r \Delta^{1/2}}{A}$$

and hence

$$\gamma = \left(\frac{-g_{33}}{\Delta} \right)^{1/2} \left(r^2 u_3^2 + A \right)^{1/2} \frac{r \Delta^{1/2}}{A}.$$

We can solve this for the speed of the particle in the ZAMO frame:

$$v^2 = \frac{\gamma^2 - 1}{\gamma^2} = \frac{r^2 u_3^2}{r^2 u_3^2 + A}. \tag{3.60}$$

Now the angular momentum of a particle in a circular orbit is given by Eq. (3.46). Multiplying numerator and denominator by r^2 this is

$$L = \pm \frac{m^{1/2} \left(r^2 \mp 2am^{1/2}r^{1/2} + a^2 \right)}{r^{3/4} \left(r^{3/2} - 3mr^{1/2} \pm 2am^{1/2} \right)^{1/2}}, \tag{3.61}$$

where the upper sign refers to prograde orbits and the lower to retrograde orbits. Now substitute Eq. (3.61) into Eq. (3.60). After some algebra this gives

$$v = \pm \frac{m^{1/2} \left(r^2 \mp 2am^{1/2}r^{1/2} + a^2 \right)}{\Delta^{1/2} \left(r^{3/2} \pm am^{1/2} \right)}. \tag{3.62}$$

We now consider various cases.

Case (1) A non-rotating hole ($a = 0$). In this case

$$v = \pm \frac{m^{1/2}}{r^{1/2} \left(1 - 2m/r \right)^{1/2}}.$$

For a particle in the last stable orbit, $r_{ms} = 6m$, we get $v = 1/2$ (or, in physical units, $v = c/2$).

Case (2) Retrograde orbits. For $a = m$ (the extreme Kerr case) the last stable retrograde circular orbit is at $r = 9m$ (section 3.13.1) for which Eq. (3.62) gives $v = -11c/26$ (in physical units). However, $r = 9m$ is outside the static limit so the energy is positive. Thus there are no retrograde circular orbits of negative energy.

Case (3) Prograde orbits. For $a = m$ the last stable prograde circular orbit is at $r = m$. To find the corresponding velocity from Eq. (3.62) we must take the limit $r \to m$ (since direct substitution gives an indeterminate form $0/0$). Putting $r = m + \delta$ and letting $\delta \to 0$ we get $v = 1/2$ (or $c/2$ in physical units). The energy of a particle in this orbit was shown to be $1/\sqrt{3}$ in section 3.13.3.

So now we know the speed of a particle in the last stable circular orbit which we are hoping will break up to yield a particle of negative energy. Consider the limiting case of a zero energy decay product. What is the speed of this particle as seen by a ZAMO? We saw in section (3.9.1) that a zero energy particle at $r = m$ is at rest with respect to a ZAMO.

Evidently a particle in the last stable orbit of an extreme Kerr black hole requires a velocity boost of $c/2$ to reduce its energy from $E = 1/\sqrt{3}$ to $E = 0$. To achieve a negative energy trajectory requires an even greater velocity boost. Bardeen et al. (1972) conclude that no plausible astrophysical process can lead to such a boost in velocity.

3.16 Spinning up a black hole

It might be thought that by allowing matter co-rotating with a black hole to be accreted the hole could be spun up indefinitely. We show below that in fact a black hole cannot be spun up beyond $a = m$ by adding mass. This is important because, as we shall see, holes with $a > m$ have some extremely undesirable properties. The following argument is from Bardeen (1970). (See also Lynden-Bell (1978).)

We consider mass falling into the hole from the last stable circular orbit. Note that any attempt to increase the angular momentum of the hole by accreting matter from other than this orbit will not help. Projecting a particle with a radial component of velocity will carry in less angular momentum and adding angular momentum to the orbit will cause the particle to escape to infinity and not be accreted at all.

When a rest mass dm_0 falls into a black hole from the last stable circular orbit the black hole acquires an increment of mass dm obtained from Eq. (3.48)

$$dm = E dm_0 = dm_0 \left(1 - \frac{2}{3} \frac{m}{r} \right)^{1/2}, \qquad (3.63)$$

with $r = r_{ms}$. It also acquires angular momentum per unit mass given by equation (3.51), so the change in the angular momentum $j = am$ of the black hole is

$$dj = L dm_0 = dm_0 \frac{2m}{3\sqrt{3}} \left[1 + 2 \left(3 \frac{r_{ms}}{m} - 2 \right)^{1/2} \right] \qquad (3.64)$$

$$= 2m \, dm \frac{x_{ms}^{1/2}}{3(3x_{ms} - 2)^{1/2}} \left[1 + 2 (3x_{ms} - 2)^{1/2} \right], \qquad (3.65)$$

where we have substituted for dm_0 from Eq. (3.63) and put $x_{ms} = r_{ms}/m$. We want to find how a/m changes so let $\alpha = a/m = j/m^2$ and return to equation (3.49), but now considered as an equation for α. Solving this quadratic in α gives

$$\alpha = \frac{1}{3} (x_{ms})^{1/2} \left[\pm 4 \mp (3x_{ms} - 2)^{1/2} \right].$$

To check the signs consider a co-rotating particle around an extreme Kerr black hole. For $x_{ms} = 1$ we get $\alpha = 1$ which is correct. For a counter-rotating particle $x_{ms} = 9$, which gives $\alpha = 1$ as it should. We have $dj = d(\alpha m^2) = m^2 d\alpha + \alpha dm^2$, which, rearranging, gives

$$m^2 \frac{d\alpha}{dm^2} = \frac{dj}{dm^2} - \alpha \qquad (3.66)$$

and using $dm^2 = 2m \, dm$ in Eq. (3.65), finally

$$m^2 \frac{d\alpha}{dm^2} = \frac{\left[1 + 2 (3x_{ms} - 2)^{1/2} \right] x_{ms}^{1/2}}{3 (3x_{ms} - 2)^{1/2}} - \alpha. \qquad (3.67)$$

   This is an implicit equation for the change in α where x_{ms} is a function of α from (3.49). It is sufficient for us to obtain our main result: if $a = m$, then $r_{ms} = m$, so $\alpha = 1$ and $x_{ms} = 1$ and we find $d\alpha/dm^2 = 0$. Thus a/m cannot increase beyond 1. The black hole can still accrete, but with $a = m$ always.

   There is a simple relationship between r_{ms} and the black hole mass as it accretes. We can eliminate α from (3.67) in favour of x_{ms} by writing

$$m^2 \frac{d\alpha}{dm^2} = m^2 \frac{d\alpha}{dx_{ms}} \frac{dx_{ms}}{dm^2}$$

and substituting for α and $d\alpha/dx_{rm}$. This gives

$$\frac{m^2}{x_{ms}} \frac{dx_{ms}}{dm^2} = -1,$$

from which

$$x_{ms} m^2 = \text{constant.} \tag{3.68}$$

Thus r_{ms} changes inversely with m. Note that equation (3.68) applies to accretion from both co-rotating and contra-rotating orbits.

Problem 59 *An extreme black hole continues to accrete from the last stable orbit, $r_{ms} = m$. Show that it remains an extreme Kerr hole with $j = m^2$.*

3.16.1   From Schwarzschild to extreme Kerr black hole

We want to look next at what happens when we start from a Schwarzschild black hole of mass m_S and spin it up by accretion of matter from the last stable circular orbit to an extreme Kerr black hole of mass m_K. The radius of the last stable orbit initially is $r_{ms} = 6m_S$ and is m_K finally. Using (3.68) we get $6m_S^2 = m_K^2$, so $m_K = \sqrt{6}m_S$ and the black hole mass has increased by $(\sqrt{6} - 1)m_S$ or 145 per cent. This is not the same as the quantity of *rest mass* that has been added. To get this we use (3.63) in the form

$$dm_0 = \frac{dm}{\left(1 - \frac{2}{3}\frac{m}{r_{ms}}\right)^{1/2}}.$$

To do the integral we have to remember that r_{ms} is a function of m; in fact, from (3.68) again , $r_{ms} = 6m_S^2/m$. Thus

$$\Delta m_0 = \int_{m_S}^{\sqrt{6}m_S} \frac{dm}{\left(1 - \frac{1}{9}\frac{m^2}{m_S^2}\right)^{1/2}}$$

$$= 3m_s \left[\sin^{-1}\left(\frac{2}{3}\right)^{1/2} - \sin^{-1}\left(\frac{1}{3}\right)^{1/2} \right],$$

or $\Delta m_0 = 1.85 m_S$. But we have shown that the black hole mass has increased by just $1.45 m_S$. Thus $0.4 m_S$ or 22 per cent of the rest mass has been radiated away in the process of accretion.

Note that the possibility of spinning up a black hole to the extreme Kerr limit depends on one crucial assumption, namely that we can neglect the effect of the radiation emitted by the accreting matter on the growth of the black hole. This is not the case. Some radiation will inevitably fall into the hole and add to its mass and angular momentum. From section 3.10.3 we see that the capture of radiation that reduces the angular momentum (counter-rotating photons) is more likely than the capture of positive angular momentum radiation. The upshot of this is that a black hole cannot be spun up to $a = m$. (In fact the limit is $a = 0.998m$ (Thorne 1974).) This outcome is comforting for two reasons. We shall see later that the structure of the spacetime of an extremal Kerr black hole is different from that of the standard Kerr model with $a < m$. Thus, in order to create an extremal hole, the global spacetime structure would have to be changed discontinuously, which is intuitively unphysical. The second reason, which will appear in section 4.4.2, is that the result that a must remain less than m generalises to a statement of the third law of thermodynamics for black holes.

Problem 60 *An extreme Kerr black hole can be spun down to a Schwarzschild black hole by accretion of contra-rotating particles from the last stable circular orbit. Show that this will increase the black hole mass by 22.5 % and that 4.7% of the accreted rest mass is radiated away in the process.*

3.17 Other coordinates

The singularity in the metric at $\Delta = 0$ is a coordinate singularity not a physical one, just like the Schwarzschild singularity at $r = 2m$. As in the Schwarzschild case the easiest way to see this is to transform to coordinates in which the metric is non-singular. The coordinates analogous to the Eddington–Finkelstein coordinates for Schwarzschild are the ingoing Kerr coordinates $(v, r, \theta, \tilde{\phi})$ and the outgoing coordinates $(u, r, \theta, \tilde{\phi})$. The ingoing Kerr coordinates are defined by

$$dv = dt + (r^2 + a^2)\frac{dr}{\Delta},$$

$$d\tilde{\phi} = d\phi + a\frac{dr}{\Delta}.$$

With these coordinates the metric becomes

$$ds^2 = \left(1 - \frac{2mr}{\rho^2}\right) dv^2 - 2dr\,dv - \rho^2 d\theta^2 - A\frac{\sin^2\theta}{\rho^2}d\tilde{\phi}^2$$

$$+ 2a\sin^2\theta\, d\tilde{\phi}\, dr + \frac{4amr}{\rho^2}\sin^2\theta\, d\tilde{\phi}\, dv,$$

which is clearly not singular at $\Delta = 0$.

Note that the transformation of t causes coordinate time v to pass increasingly quickly as we approach the horizon, which, just as in Schwarzschild, counteracts the 'stalling' of the infalling trajectory there. But in addition the transformation in ϕ 'unwinds' the angular coordinate increasingly rapidly as the horizon is approached to counteract the infinite winding of the infalling trajectory that we found in section 3.8.1. One further difference from the Schwarzschild case is that the lack of spherical symmetry means that the picture we draw, with the angle θ suppressed, is different for different values of θ. Note also however, that most geodesics (and hence light rays) do not have constant values of θ, so in general a figure with θ suppressed does not give a very realistic view of the spacetime. (It does not allow one to picture the free fall particle paths.) Exceptions occur in the equatorial plane $(\theta = \pi/2)$ and for polar orbits.

As for Schwarzschild, to draw the space-time diagram we define a new time $\tilde{t} = v - r$, and plot the trajectories of light rays in $(\tilde{t}, r, \phi)$ space. This is shown in Fig. 4 for the case $\theta = \pi/2$. It is clear from the diagram why we get two special null surfaces: on the static limit surface the counter-rotating null lines have been 'unwound' by the rotation of the hole as the 'straight' null lines that generate the surface, while on the horizon the outgoing light rays are wound up to remain in the surface.

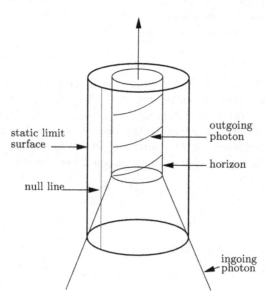

Figure 4 Photons around a Kerr black hole (based on Frolov and Novikov p. 79).

3.18 Penrose–Carter diagram

In the case of a Schwarzschild black hole we obtained the continuation of the geometry into the interior of the horizon by transforming to Kruskal coordinates. We can do a similar thing for the Kerr solution. It turns out that there are three very different cases to consider. There is also an important difference in relating the pure vacuum solutions to the physical collapse of a star, but we shall come to that in section 3.18.1. The three cases are $a > m$, which will turn out to be the simplest, but is quite unphysical in interesting ways; $a = m$, the (unattainable) limiting case referred to as the extreme (or, sometimes, maximal) Kerr solution; and the apparently most complicated case $a < m$. (The following discussion is adapted from Townsend (unpublished).)

(i) $a > m$. This case is simple enough for us to provide the algebraic details. The metric has a coordinate singularity when $\Delta = (r - r_+)(r - r_-) = 0$, where $r_\pm = m \pm (m^2 - a^2)^{1/2}$. Thus for $a > m, \Delta = 0$ has no real solutions and there are no coordinate singularities (except the usual one at $\theta = 0$). It can be shown that there is a real singularity at $\rho = 0$, i.e. at $r = 0$, $\theta = \pi/2$. To see the structure of the spacetime in this case we can use a further set of coordinates, the Kerr-Schild coordinates (T, X, Y, Z). These are defined by

$$X = (r \cos \tilde{\phi} + a \sin \tilde{\phi}) \sin \theta$$
$$Y = (r \sin \tilde{\phi} - a \cos \tilde{\phi}) \sin \theta$$
$$Z = r \cos \theta$$
$$dT = dt - \left(\frac{r^2 + a^2}{\Delta} - 1\right) dr,$$

where $\tilde{\phi}$ is, as defined above,

$$\tilde{\phi} = \phi - \int a \frac{dr}{\Delta}.$$

and it follows that r is given implicitly as a function of (X, Y, Z) by

$$r^4 - (X^2 + Y^2 + Z^2 - a^2)r^2 - a^2 Z^2 = 0. \tag{3.69}$$

The metric in Kerr-Schild coordinates becomes

$$ds^2 = dT^2 - dX^2 - dY^2 - dZ^2$$
$$- \frac{2mr^3}{r^4 + a^2 Z^2} \left[\frac{-r(XdX + YdY) - a(XdY - YdX)}{r^2 + a^2} - \frac{ZdZ}{r} + dT\right]^2.$$

At $r = 0$ we have $Z = 0$. Also if $r = 0$, then $dr = 0$, and if $\theta = \pi/2$ we get $X = a \sin \tilde{\phi}$, $Y = a \cos \tilde{\phi}$, so $X^2 + Y^2 = a^2$. Thus the metric is singular on this ring in the (X, Y) plane (i.e. the equatorial plane). All inward null rays in the equatorial plane must intersect this ring.

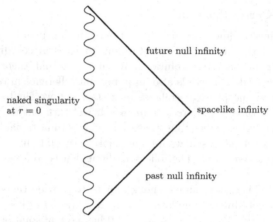

Figure 5   Kerr spacetime for $a > m$ (based on Townsend, p. 81).

To construct a picture of the spacetime for $\theta = \pi/2$ we imagine a further coordinate transformation that has no effect near $r = 0$ but brings the asymptotically flat region at large r in to finite values, and make a conformal transformation to a metric that is non-singular at the outer boundaries. This will result in the picture of Fig. 5. Since there is no horizon, light rays from the ring singularity escape to future null infinity, and the ring is visible to external observers. The ring is therefore a naked singularity. According to the cosmic censorship hypothesis (section 2.12.1) this should mean that the Kerr solution with $a > m$ is unphysical (i.e. the gravitational field cannot be created from physical matter).

Recall that we should not be misled by the choice of symbols for coordinates into ascribing conventional meanings to them in relativity. We have here another case in point: the metric gives a valid geometry for negative values of the coordinate r. The spacetime therefore continues through the disc $X^2 + Y^2 = a^2$ to another asymptotically flat region with negative r. The whole spacetime picture in the equatorial plane therefore contains another wedge to the left of the ring $r = 0$ in Fig. 5.

For other values of θ, $r = 0$ corresponds to points in the disc $X^2 + Y^2 < a^2$ interior to the ring. Null geodesics pass through the disc to the $r < 0$ region.

(ii) $a < m$. The analogy between the Kerr ingoing and outgoing coordinates and the two sets of Eddington–Finkelstein coordinates suggests that we can find the analogue of the Kruskal extension of Schwarzschild for the Kerr metric. This is indeed the case. It turns out that the algebraic details are not very illuminating (see e.g. Frolov and Novikov, appendix) so we just give the result in the form of the Penrose–Carter diagram in the equatorial plane (Fig. 6). This contains the two asymptotically flat regions I and IV, and the past and future horizons at $r = r_+$, as we might expect from the analogy. The regions interior to the horizons are quite

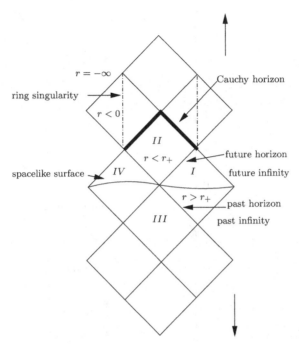

$r = -\infty$

ring singularity

$r < 0$

Cauchy horizon

II

$r < r_+$

future horizon

spacelike surface

IV

I

future infinity

$r > r_+$

past horizon

III

past infinity

Figure 6 Penrose–Carter diagram of the Kerr solution with $a < m$. The figure is repeated infinitely in both directions.

surprising however. Regions II and III are bounded not by a spacelike singularity (a singularity where time-like geodesics come to an end, as in the Kruskal metric) but by the inner horizons at $r = r_-$. These horizons act as limits to predictability (called Cauchy horizons) for exterior observers. To understand this, consider that the positions and velocities of all particles in the exterior of the black hole are given on a spacelike surface in region I. The positions of these particles can be predicted up to the Cauchy horizon. Beyond that a new particle could come in from another region and collide with the given particles thereby upsetting any further future predictions.

Within these Cauchy horizons are the ring singularities, which are timelike singularities since they terminate spacelike geodesics. (Such singularities are avoidable in contrast to spacelike singularities.) There are also further asymptotically flat regions with negative values of r, extending to $r = -\infty$. And then the whole pattern is repeated indefinitely to the past and to the future! Before commenting we look finally at the extreme Kerr case.

(iii) $a = m$. This is similar to $a < m$, except that the two horizons at $r = r_\pm = m$ coincide, so the region of spacetime between them is no longer present (Fig. 7).

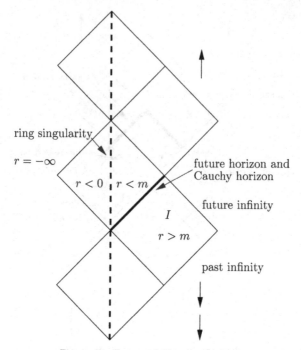

Figure 7   Extremal Kerr spacetime.

Two comments are in order. The first is the question of what happens if we try to spin up a black hole by adding angular momentum. If the case $a > m$ is unphysical, there must be some limit to the accretion of angular momentum. We explored this in section 3.16 and found that there is a limit. The second is the nature of the interior solution, and especially the possibility of exploring all those infinity of separate regions ('universes') within the horizon of a spinning black hole. One should be slightly suspicious of the physical reality of this picture, since the multiplicity of universes would appear to be created as soon as a Schwarzschild hole acquires the tiniest amount of angular momentum. We shall deal with this (as far as possible) in the next section.

3.18.1   Interior solutions and collapsing stars

We have emphasised that the Kerr solution, unlike the Schwarzschild solution, is not the metric of the vacuum spacetime exterior to a material body. Furthermore, in the collapse of a spherical body, the metric outside the body, both outside *and inside* the horizon, is exactly the Schwarzschild solution. Real stars are not exactly spherical, but if they are non-rotating the Schwarzschild solution is an increasingly good approximation to the metric as the collapse proceeds. This is not the case for the collapse of a rotating body: inside the horizon the metric is not even approximately the Kerr solution because the Kerr metric does not incorporate the gravitational waves that would be produced in a real collapse. This means that the exotic properties of the Kerr solution interior to the horizon do not represent the situation in a black hole that is formed by collapsing matter.

Suppose however that we were to be presented with an exact Kerr black hole. Could we explore the many universes of the interior? Probably not. Any observer exploring this metric represents a perturbation of the geometry. Near the inner horizon this perturbation cannot remain small, so the spacetime, including the observer, cannot be approximated by the exact Kerr black hole. In this sense the exact Kerr black hole interior cannot be explored without destroying it.

3.19 Closed timelike lines

The Kerr metric exhibits another interesting pathology. It contains closed time-like curves along which an observer would appear to be able to travel into his or her past. We show this first, and then discuss its significance.

Consider an orbit along which only ϕ is changing. The proper time along this orbit is given by

$$d\tau^2 = g_{33}d\phi^2$$
$$= \left[-a^2 \sin^2\theta \left(1 + \frac{r^2}{a^2}\right) - \frac{ma^2}{r}\left(\frac{2\sin^4\theta}{1 + \frac{a^2}{r^2}\cos^2\theta}\right)\right]d\phi^2.$$

Usually this quantity would be negative indicating a spacelike interval. We shall show there are curves along which this $d\tau^2$ is positive. Since curves in ϕ are closed, these will be closed timelike lines. Put $\theta = \pi/2 + \delta$ and $r = a\delta$, with δ small. Then

$$d\tau^2 = \left[-a^2 + \cdots - \frac{ma}{\delta} + \cdots\right]d\phi^2,$$

where we have indicated the omission of terms in δ and higher powers. If $\delta < 0$ and small then clearly $d\tau^2$ is positive as promised. Thus, in the region $r < 0$ there are closed timelike paths. (Note that these are not geodesics, so some effort is required to navigate them.)

Problem 61 *The existence of closed timelike lines can also be demonstrated in the Kerr-Schild coordinates. Show this by calculating the element of proper time along a circle $r = a\delta$, $Z = $ constant, at constant time T. [Hint: along a circle in the (X, Y) plane $dX \propto -Y$, $dY \propto X$.]*

The problems with being able to travel into the past are obvious and potentially gruesome. (Add this to the Oedipus legend, for example.) Fortunately in this case the cosmic censorship hypothesis (if true) would prevent such a thing. In fact, as we have seen, we do not know how to make a Kerr black hole with $a > m$ using normal matter. On the other hand, travel into the past through singularity-free wormholes does appear to be possible (in principle) according to current theories (see chapter 5). Whether more complete understanding of physics (including quantum gravity for example) will turn out in practice to allow only consistent time-travel or to prevent it altogether remains to be seen.

3.20   Charged black holes

Generally speaking astronomical bodies are electrically neutral, but there are several reasons why the addition of electric charge to black holes has been considered. The first is to add to the bank of exact solutions that can have their properties explored. In this respect charged (non-rotating) black holes act as a bridge to the study of Kerr black holes with which they have in common a Cauchy horizon and a timelike singularity, but lack some of the technical complexity. The second is one of completeness: the only external parameters black holes can have are mass, angular momentum and electrical charge (neglecting a hypothetical magnetic charge), so the Kerr-Newman metric of a charged spinning black hole is the most general black hole solution. The lack of other conserved quantities beyond mass, charge and angular momentum, associated with black holes is often expressed by saying that 'black holes have no hair'. One other importance that the extremal charged black hole has acquired in recent times is that it provides a connection to string theory.

The special case of a spherical charged black hole is known as the Reissner-Nordström solution. This has the metric

$$dr^2 = \left(1 - \frac{2m}{r} + \frac{q^2}{r^2}\right) dt^2 - \frac{dr^2}{\left(1 - \frac{2m}{r} + \frac{q^2}{r^2}\right)} - d\Omega^2.$$

Now $g_{00} = 0$ when

$$r_\pm = m \pm \left(m^2 - q^2\right)^{1/2}.$$

If $|q| > m$ then the roots of the quadratic are not real so g_{00} stays positive at all values of r except at $r = 0$, where there is a singularity.

The more interesting case is when $|q| < m$. There are three regions: (I) $r_+ < r < \infty$; (II) $r_- < r < r_+$; (III) $0 < r < r_-$. In region (I) g_{00} is positive so it is possible to remain stationary. In region (II) let $r = r_+ - \delta$. Then

$$g_{00} = \left[1 - \frac{2m}{r_+}\left(1 - \frac{\delta}{r_+}\right)^{-1} + \frac{q^2}{r_+^2}\left(1 - \frac{\delta}{r_+}\right)^{-2}\right]$$

$$= -\frac{2\delta}{r_+^2}\left(m - \frac{q^2}{r_+}\right).$$

Now, $q < m$ and $r_+ > m$ so $g_{00} < 0$ in (II). Thus in region (II) a stationary particle would have a spacelike world line so rest is not possible. From the figure we see that particles can cross $r = r_+$ in one direction only. Thus $r = r_+$ is a horizon. In region (III), $r = r_- - \delta$ and

$$g_{00} = -\frac{2\delta}{r_-^2}\left(m - \frac{q^2}{r_-}\right).$$

But now $r_- < m$ and $q^2 < m^2$ so $g_{00} > 0$. Figure 9 shows a plot of g_{00} against r. In region (III) it is possible to remain stationary and avoid the singularity or to move into another region (see Fig. 8). The fate of a particle falling into a charged black hole is more interesting than that of one falling into a Schwarzschild hole! The singularity is timelike, in contrast to the spacelike singularity in the Schwarzschild metric, which cannot be avoided.

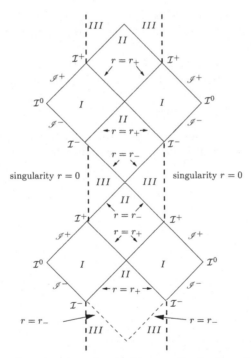

Figure 8   Penrose–Carter diagram for the Riessner–Nordström solution with $q < m$.

Finally, for $q = m$, (the extremal case) we have $r_+ = r_- = m$ and $g_{00} = 0$ at $r = r_\pm$. There is no region (II) in this case.

The Kerr-Newman metric in Boyer-Lindquist coordinates for a spinning black

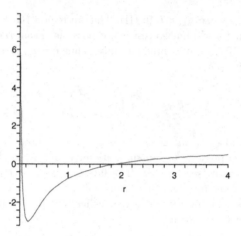

Figure 9 g_{00} versus radius r in the case $m = 1$, $q = 0.5$. Note that within the inner horizon g_{00} becomes positive so the time coordinate is timelike as is the singularity at $r = 0$.

hole with charge q is obtained from the Kerr metric by the replacement

$$2mr \to 2mr - q^2.$$

The properties of this solution can be explored in a similar manner to those of the Kerr metric. In particular the event horizon is given by $\Delta = r^2 + a^2 - 2mr + q^2 = 0$ and exists only for $m^2 \geq a^2 + q^2$.

Problem 62 *Show that the Kerr-Newman metric has an event horizon only in the case $m^2 \geq a^2 + q^2$.*

Chapter 4

BLACK HOLE THERMODYNAMICS

In this chapter we introduce the thermodynamics of black holes. Our strategy is to begin by establishing some mathematical identities for the Kerr solution (and, a fortiori, the Schwarzschild solution). These mathematical identities have a resemblance to the laws of thermodynamics if one assigns a temperature to a black hole. This resemblance was initially regarded as a curiosity, because a black hole did not appear to behave like a body with a temperature in one important respect: it did not radiate. In fact, the area theorem, which we introduce in section 4.2, appears to prevent the emission of radiation from a Kerr black hole. However, once quantum theory is taken into account, Hawking showed that black holes do emit radiation precisely as required by the thermodynamical interpretation. Although it is beyond our scope to show this, these results can be extended to hold not just for Kerr black holes but for black holes in general. Thus black holes do obey the laws of thermodynamics.

4.1 Black hole mechanics

4.1.1 Surface gravity

In section 2.8.2 we calculated the surface gravity of a Schwarzschild black hole by looking at the force on a rope supporting a body just above the horizon. In Kerr spacetime we cannot use this physical picture, because any part of the rope that was lowered into the ergosphere would be torn off by the rotation, since nothing can remain stationary inside the static limit surface. But we can obtain the acceleration of a zero angular momentum observer as seen from infinity in a similar way.

We begin by computing the local acceleration of a zero angular momentum observer. From Eq. (3.37) such an observer has the 4-velocity

$$u^\mu_{\text{ZAMO}} = \left[\left(\frac{-g_{33}}{\Delta \sin^2\theta} \right)^{1/2}, 0, 0, \omega \left(\frac{-g_{33}}{\Delta \sin^2\theta} \right)^{1/2} \right] \tag{4.1}$$

$$= \left[\left(\frac{A}{\rho^2\Delta} \right)^{1/2}, 0, 0, \frac{2mra}{A} \left(\frac{A}{\rho^2\Delta} \right)^{1/2} \right],$$

and hence the 4-acceleration

$$a^\mu = \frac{du^\mu_{\text{ZAMO}}}{d\tau} + \Gamma^\mu_{\nu\sigma} u^\nu_{\text{ZAMO}} u^\sigma_{\text{ZAMO}}. \tag{4.2}$$

The only non-zero contribution comes from the radial acceleration a^1. So Eq. (4.2) reduces to

$$a^1 = \left(u^0_{\text{ZAMO}}\right)^2 \left(\Gamma^1_{00} + 2\Gamma^1_{03}\omega + \Gamma^1_{33}\omega^2\right),$$

where $u^3/u^0 = \omega$ as the observer has zero angular momentum. Substituting for ω from Eq. (3.28) gives

$$a^1 = \frac{\left(u^0_{\text{ZAMO}}\right)^2}{A^2} \left(A^2\Gamma^1_{00} + 4mar A\Gamma^1_{03} + 4m^2a^2r^2\Gamma^1_{33}\right). \tag{4.3}$$

We therefore need the components Γ^1_{00}, Γ^1_{30} and Γ^1_{33} of the affine connection. These can be calculated directly from Eq. (1.10) and the Boyer-Lindquist form of the metric. We obtain

$$\Gamma^1_{00} = -\frac{g^{11}m}{\rho^4}\left(r^2 - a^2\cos^2\theta\right),$$

$$\Gamma^1_{30} = \frac{g^{11}m}{\rho^4}a\sin^2\theta\left(r^2 - a^2\cos^2\theta\right),$$

$$\Gamma^1_{33} = \frac{g^{11}}{2\rho^4}\sin^2\theta\left(\rho^2\frac{dA}{dr} - 2Ar\right).$$

From Eq. (1.13) the proper acceleration at coordinate radius r is given by

$$g_r^2 = -a^\mu a_\mu = -g_{11}\left(a^1\right)^2.$$

Substituting into Eq. (4.3) for the connection coefficients and taking out the common factor $-mg^{11}/\rho^4$ gives

$$g_r = \frac{m}{\rho^7\Delta^{1/2}A}F$$

where

$$F = \left[A^2\left(r^2 - a^2\cos^2\theta\right) - 4Ama^2r\sin^2\theta\left(r^2 - a^2\cos^2\theta\right) - 2ma^2r^2\sin^2\theta\left(A'\rho^2 - 2Ar\right)\right]$$

and $A' = dA/dr$. As we approach the horizon the locally measured acceleration $g_r \to \infty$. However, the acceleration g_∞ measured from infinity is finite since the two accelerations are related by a redshift factor as we explained in section 2.8.2. Specifically

$$g_\infty = g_r\frac{E_\infty}{E_r}$$

(compare Eq. (2.8.2)). We compute $E_r/E_\infty = (A/\rho^2\Delta)^{1/2}$ in the next section, giving

$$g_\infty = \frac{m}{\rho^6 A^{3/2}}F.$$

On the horizon, $\Delta = 0$, $r = r_+ = m + (m^2 - a^2)^{1/2}$, so

$$A = (r_+^2 + a^2)^2 = 4m^2r_+^2.$$

Inserting these values into F gives, after some algebra,

$$F(r_+, \theta) = 4mr_+^2(r_+ - m)\rho^6.$$

F is a function of θ through ρ, but the ρ^6 dependence will cancel in g_∞, leaving g_∞ independent of θ. Specifically, by definition, the acceleration on the horizon measured from infinity is $\kappa = g_\infty(r_+)$. So, using $(r_+ - m) = (m^2 - a^2)^{1/2}$, we get finally,

$$\kappa = \frac{(m^2 - a^2)^{1/2}}{2mr_+}. \tag{4.4}$$

In Schwarzschild spacetime the surface gravity must be constant on the horizon by virtue of the spherical symmetry. Our expression for surface gravity in the Kerr metric shows it to be constant on the horizon in this case also.

Note that for an extreme Kerr hole, $a = m$, we have $\kappa = 0$. This says that the surface gravity is zero because the rotation balances gravity. It is equivalent to a normal body rotating at break-up speed.

Problem 63 *(i) Show that the vector field $(\chi^\mu) = (1, 0, 0, \omega)$ is null on the horizon. (ii) More sophisticated treatments define the surface gravity in terms of the rate of change of the null vector field $(\chi^\mu) = (1, 0, 0, \omega)$ through*

$$\chi^\nu(\nabla_\nu \chi^\mu) = \kappa \chi^\mu.$$

Use the ingoing Eddington–Finkelstein coordinates (which are non-singular on the horizon) to derive the surface gravity of a Schwarzschild hole, $\kappa = 1/4m$, from this definition.

The result Eq. (4.4) for κ can be obtained for a Kerr black hole using the ingoing Kerr coordinates introduced in section 3.17.

4.1.2 Redshift

To complete the derivation of the expression for surface gravity obtained in the last section we need to derive the expression for the red shift, at infinity, of light emitted from a zero angular momentum observer at radial coordinate r. The time component of the momentum 4-vector of a zero angular momentum photon is $p_0 = E_{ph}$ where E_{ph} is the conserved energy of the photon measured by an observer at infinity. Also $(L_{ph})_z = 0$ so $p_3 = 0$. Now we can obtain a relation between the energy of a photon measured locally at coordinate r by a ZAMO and the energy of the photon measured at infinity. This we do by evaluating the scalar product $p_\mu u^\mu_{\text{ZAMO}}$ in the local frame and in the original Kerr frame and equating the results. So

$$E_r = E_{\text{ph}} \left(\frac{-g_{33}}{\Delta \sin^2 \theta} \right)^{1/2},$$

where, as usual, the components of u^μ_{ZAMO} in the Schwarzschild frame are given by equation (4.1) and the components of $(p_\mu) = (E_{\text{ph}}, p_1, p_2, 0)$.

The conserved energy E_{ph} is the photon energy measured at infinity, $E_\infty = E_{\text{ph}}$ and therefore, using the identities in section 3.3.1,

$$\frac{E_r}{E_\infty} = \left(\frac{A}{\rho^2 \Delta}\right)^{1/2}.$$

4.1.3   Conservation of energy

We now look at the energy change, or equivalently the change in mass, of a Kerr black hole under an external perturbation, such as the accretion of an element of mass. This will allow us to derive a relation expressing the conservation of energy for the black hole (not to be confused with the conservation of energy of a test body moving in the Kerr metric).

Recall from section 3.2.2 (Eq. (3.9)) that the area of the event horizon is

$$A_{\text{h}} = 8\pi m \left[m + (m^2 - a^2)^{1/2}\right]. \tag{4.5}$$

A change in A_{h} is related to a change in m and a by

$$\delta A_{\text{h}} = 8\pi[2m\delta m + \delta m(m^2 - a^2)^{1/2} + m(m^2 - a^2)^{-1/2}(m\delta m - a\delta a)], \tag{4.6}$$

or

$$\delta A_{\text{h}} = \frac{8\pi}{(m^2 - a^2)^{1/2}} \left[2mr_+\delta m - a^2\delta m - am\delta a\right].$$

The angular momentum of the black hole is $j = ma$, so $\delta j = a\delta m + m\delta a$ and hence

$$\delta A_{\text{h}} = \frac{16mr_+\pi}{(m^2 - a^2)^{1/2}} \left[\delta m - \omega_+\delta j\right], \tag{4.7}$$

where $\omega_+ = a/(2mr_+)$ is the angular velocity of the black hole. We can rewrite this as

$$\delta m = \frac{\kappa}{8\pi}\delta A_{\text{h}} + \omega_+\delta j, \tag{4.8}$$

where

$$\kappa = (m^2 - a^2)^{1/2}/(2mr_+)$$

is the surface gravity. Eq. (4.8), which expresses the conservation of energy, is the first law of black hole mechanics.

Problem 64 *Smarr showed that Eq. (4.8) could be integrated to obtain*

$$m = \frac{\kappa}{4\pi}A_{\text{h}} + 2\omega_+j.$$

Verify this for the Kerr metric by explicit calculation.

4.2 The area of a Kerr black hole horizon cannot decrease

We expect that the addition of a particle of mass δm with zero angular momentum to a black hole will increase the mass by δm, so according to Eq. (4.7), the area of the event horizon will increase on addition of mass. Conversely, if we could add angular momentum without adding mass then Eq. (4.7) would show that the area would decrease. We certainly could not add angular momentum without mass by firing particles into the hole, so we would need to be more inventive. We look first at what happens to the surface area if we accrete matter from the last stable orbit. We shall then investigate what happens if energy and angular momentum are extracted from a black hole by the Penrose process. In this case δm and δj are negative, so does the minus sign in Eq. (3.64) come to the rescue and lead to an increase in area? In this example we can show that indeed it does. These examples illustrate the general result that the area of the horizon cannot decrease.

4.2.1 Area change by accretion

We consider accretion from the last stable orbit of an extreme Kerr black hole. From Eq. (3.48) the accretion of a mass δm_0 from the last stable prograde orbit at $r_{ms} = m$ gives rise to a change in mass of the hole

$$\delta m = \delta m_0 \left(1 - \frac{2}{3}\frac{m}{r_{ms}}\right)^{1/2} = \frac{\delta m_0}{\sqrt{3}}.$$

From Eq. (3.51) the associated change in angular momentum is

$$\delta j = \delta m_0 \frac{2m}{3\sqrt{3}} \left[1 + 2\left(3\frac{r_{ms}}{m} - 2\right)^{1/2}\right].$$

With $r_{ms} = m$ this becomes

$$\delta j = \frac{2m\delta m_0}{\sqrt{3}}.$$

So, since $\omega_+ = a/(2mr_+) = 1/2m$ for $a = m$ and $r_+ = m$, we get

$$\delta m - \omega_+\delta j = 0.$$

Direct substitution in Eq. (4.7) for $a = m$ would give the indeterminate form $0/0$. Therefore we have to evaluate the limit $a \to m$ more carefully.

Let $\varepsilon = m - a$ be small and positive. Then

$$\delta m = \frac{\delta m_0}{\sqrt{3}} + O(\varepsilon) \quad \text{and} \quad \delta j = \frac{2m\delta m_0}{\sqrt{3}} + O(\varepsilon).$$

Now substitute for δm, δj, and ω_+ into equation (4.7). This gives

$$\delta A_{\mathrm{h}} = \frac{16\pi r_+ m}{(m^2 - a^2)^{1/2}} \left(\frac{\delta m_0}{\sqrt{3}} - \frac{a}{r_+}\frac{\delta m_0}{\sqrt{3}} + O(\varepsilon)\right).$$

But

$$r_+ = m + (m^2 - a^2)^{1/2}$$

so

$$r_+ - a = m - a + (m^2 - a^2)^{1/2} = \varepsilon + (m^2 - a^2)^{1/2}.$$

Using this result gives

$$\delta A_{\mathrm{h}} = \frac{16\pi m}{(m^2 - a^2)^{1/2}} \frac{\delta m_0}{\sqrt{3}} (r_+ - a + O(\varepsilon))$$

$$\to \frac{16\pi}{\sqrt{3}} m \delta m_0 > 0.$$

Thus the horizon area does indeed increase in this case.

Problem 65 *Show that accretion of a photon from the circular photon orbit at radius* r_{ph} *also increases the area of an extreme Kerr black hole.*

4.2.2   Area change produced by the Penrose process

In the Penrose process a particle of energy E_0 inside the ergosphere decays into a positive energy particle of energy E_P and a negative energy particle of energy E_N, where $E_0 = E_N + E_P$. The positive energy particle escapes to infinity and the negative energy particle is captured by the black hole with a change (loss) of mass (or energy)

$$\delta m = E_N. \tag{4.9}$$

Similarly, since the negative energy particle has negative angular momentum, the hole loses angular momentum

$$\delta j = L_N. \tag{4.10}$$

The desired result hinges on the fact that $|E_N| \leq \omega_+ |L_N|$, which we derived in section 3.15.2. Using this result we get, from Eq. (4.9) and Eq. (4.10)

$$|\delta m| \leq \omega_+ |\delta j|.$$

From Eq. (4.7), remembering that both δj and δm are negative, the area law is

$$\delta A_{\mathrm{h}} = \frac{8\pi}{\kappa} (\delta m - \omega_+ \delta j) \geq 0.$$

A transformation can be reversed only if $\delta A_{\mathrm{h}} = 0$; if $\delta A_{\mathrm{h}} > 0$ the transformation is obviously irreversible (because the inverse transformation would violate the area law).

4.2.3 The area theorem

We have demonstrated the increase in area for particular processes and specifically for a Kerr black hole. Hawking showed that the result holds within classical general relativity, under certain technical assumptions, for any process and any black hole (not just those in a vacuum). This result is known as the Hawking area theorem, or the second law of black hole mechanics:

No interaction whatsoever can result in a decrease in the total surface area of a (classical) black hole.

4.2.4 Irreducible mass

Equation (4.5) can be written as

$$m^2 = \frac{A_h}{16\pi} + \frac{4\pi j^2}{A_h}. \tag{4.11}$$

If we reduce j to zero reversibly then $\delta A_h = 0$ giving the smallest value for m^2. This is called the irreducible mass m_{ir}, so we have

$$m_{ir}^2 = \frac{A_h}{16\pi}$$

or

$$A_h = 4\pi \left(\frac{2GM_{ir}}{c^2} \right)^2,$$

analogous to the expression in the Schwarzschild case, where M_{ir} is the irreducible mass in physical units.

We have therefore

$$m^2 = m_{ir}^2 + \frac{4\pi j^2}{A_h}.$$

The second term on the right is the contribution to the mass (or energy) of the hole coming from the rotation. This is another version of the energy equation (or the Smarr formula or the first law) Evidently 'the Schwarzschild black hole is the ground state of the Kerr black hole'.

We shall look at some consequences of the 'area theorem' (that the area of a black hole can never decrease) below.

Problem 66 Show that Eq. (4.5) can be written in the form of Eq. (4.11).

Problem 67 Show that for a slowly rotating black hole the rotational energy can be expressed as $m_{rot} = \frac{1}{2}I\omega_+^2$ where I, the moment of inertia, is $4m^3$.

4.2.5 Maximum energy extraction

We can use the above results to obtain a limit to the amount of energy that can be extracted from a spinning black hole. This is $m - m_{ir}$. But

$$m_{ir}^2 = \frac{A_H}{16\pi} = \frac{m}{2}\left[m + (m^2 - a^2)^{1/2}\right].$$

Now m_{ir} is a minimum, hence $m - m_{ir}$ is a maximum when $a = m$, and therefore for an extreme Kerr black hole we get

$$m - m_{ir} = m - \frac{m}{\sqrt{2}}.$$

The maximum energy that can be extracted is thus 29 per cent of the mass of the black hole.

Problem 68 *Show that the maximum energy that can be extracted from the merging of two Schwarzschild black holes of equal mass to form a single black hole is 29 per cent of the original mass.*

Problem 69 *Two extreme Kerr black holes of the same mass and equal but opposite angular momentum coalesce reversibly to form a Schwarzschild black hole. How much energy is radiated away in the process?*

4.2.6 Naked singularities

So what happens if we try to add mass and angular momentum to an extreme Kerr black hole? Since the area cannot decrease, we have, from Eq. (4.6),

$$\delta A_h = 8\pi\delta m\left[m + (m^2 - a^2)^{1/2}\right] + 8\pi m\left[\delta m + (m^2 - a^2)^{-1/2}(m\delta m - a\delta a)\right] \geq 0.$$

This can be re-arranged to give

$$\left[2m(m^2 - a^2)^{1/2} + 2m^2 - a^2\right]\delta m \geq ma\delta a.$$

For $a \to m$ this becomes

$$m\delta m \geq a\delta a,$$

hence, integrating, $m^2 \geq a^2$, or, equivalently, $a \leq m$. The area theorem prevents the spinning up of a black hole beyond the extremal state. In other words, it prevents the formation of a naked singularity by this means. However, it appears to allow the extreme Kerr black hole with $a = m$ to be reached. In fact, we have seen in section 3.16.1 that, if we go beyond the accretion of test matter to take into account all the physics of the accretion process, then this is impossible. The inability to reach $a = m$ in a finite number of physical operations is an example of the third law of black hole thermodynamics (section 4.4.2).

4.3 Scattering of waves

4.3.1 Superradiance

So far we have looked at the scattering of particles by a black hole. In order to approach the quantum theory of black holes it will be of interest to consider how waves are scattered by a black hole. We shall confine ourselves to a brief outline. Details can be found in Frolov and Novikov (1998) or Futterman *et al* (1988) and the references therein.

To simplify the discussion we consider a (hypothetical) scalar wave $\psi(t, r, \theta, \phi)$ for a massless field, but the conclusions we reach apply equally to the electromagnetic field. The massless scalar wave satisfies the covariant form of the wave equation

$$\nabla_\mu \nabla^\mu \psi = 0. \tag{4.12}$$

As can be verified by explicit calculation using Eq. (1.10), this can be written as

$$(-g)^{-1/2} \partial_\mu \left[(-g)^{1/2} g^{\mu\nu} \partial_\nu \psi \right] = 0, \tag{4.13}$$

where, as usual, $\partial_\mu \equiv \partial / \partial x^\mu$ and $g = \det(g_{\mu\nu})$. By taking $g_{\mu\nu}$ to be those of the spherically symmetric Schwarzschild metric, the equation can be written explicitly in (t, r, θ, ϕ) coordinates

$$\frac{\partial^2 \psi}{\partial t^2} - \frac{2}{r} \left(1 - \frac{m}{r} \right) \left(1 - \frac{2m}{r} \right) \frac{\partial \psi}{\partial r} - \left(1 - \frac{2m}{r} \right)^2 \frac{\partial^2 \psi}{\partial r^2}$$
$$- \left(1 - \frac{2m}{r} \right) \frac{1}{r^2 \sin^2 \theta} \frac{\partial}{\partial \theta} \left(\sin \theta \frac{\partial \psi}{\partial \theta} \right) - \left(1 - \frac{2m}{r} \right) \frac{\partial^2 \psi}{\partial \phi^2} = 0.$$

As can be readily verified by substitution, this has separable solutions of the form,

$$\psi(t, r, \theta, \phi) = R_{\omega L}(r) Y_{LM}(\theta, \phi) e^{\pm i\omega t}, \tag{4.14}$$

where L, M and ω are separation constants. The $Y_{LM}(\theta, \phi)$ are the usual spherical harmonics satisfying the angular part of the wave equation, which has the same form as in flat spacetime. A general solution for ψ would be made up of a linear superposition of these separable solutions with arbitrary coefficients, as usual.

The radial function $R_{\omega L}(r)$ satisfies

$$\left(1 - \frac{2m}{r} \right)^2 \frac{d^2 R_{\omega L}}{dr^2} + \frac{2}{r} \left(1 - \frac{m}{r} \right) \left(1 - \frac{2m}{r} \right) \frac{dR_{\omega L}}{dr}$$
$$+ \left[\omega^2 - \left(1 - \frac{2m}{r} \right) \frac{l(l+1)}{r^2} \right] R_{\omega L} = 0. \tag{4.15}$$

We now introduce the function

$$u_{\omega L}(r) = r R_{\omega L}(r)$$

and the 'tortoise'coordinate, Eq. (2.45),

$$r_* = r + 2m \log \left(\frac{r}{2m} - 1 \right).$$ (4.16)

Since it is obvious from Eq. (4.15) that $u_{\omega L}(r)$ depends on the parameters ω and L we drop the subscripts for clarity. The function $u = u_{\omega L}(r)$ then satisfies the Regge-Wheeler equation

$$\frac{d^2u}{dr_*^2} + \omega^2 u - V_L(r)u = 0,$$ (4.17)

where V_L is the effective potential

$$V_L(r) = \left(1 - \frac{2m}{r} \right) \left[\frac{l(l+1)}{r^2} + \frac{2m}{r^3} \right],$$

which is implicitly a function of r_* through (4.16).

Note that $r_* \to +\infty$ as $r \to \infty$ and $r_* \to -\infty$ as $r \to 2m$, so r_* covers the exterior of the black hole. Note also that, since the effective potential tends to zero on the horizon and at infinity, the solutions of (4.17) at infinity and at the horizon have the simple form in terms of r_*

$$u \sim e^{\pm i\omega r_*}.$$

The simple behaviour in terms of r_*, rather than r, reflects the long range behaviour of the gravitational force.

Consider the solution of the wave equation in which no waves emerge from the black hole. This requires that we include in ψ only ingoing modes at $r_* \to -\infty$, hence these solutions are often called the 'in' modes. As in the standard case of the one-dimensional wave equation (for example for waves on a string), this requires that ψ be a function of $r_* + t$ (and not $r_* - t$). From the form of ψ in Eq. (4.14) we see that we must have

$$u \sim e^{-i\omega r_*} \quad \text{as } r_* \to -\infty.$$ (4.18)

Since ω is positive by definition, from a quantum mechanical viewpoint we can also see that this wavefunction has negative momentum $-\hbar\omega$ (since $-i\hbar\partial u/\partial r_* = -\hbar\omega u$) and hence represents a wave travelling to smaller r (or r_*).

Just as for the standard scattering of waves in one dimension, having set the conditions at $r_* \to -\infty$ we can in principle calculate from the wave equation what happens as $r_* \to +\infty$. The ingoing wave will be scattered by the black hole such that part goes down the hole and part is reflected back to future infinity (Fig. 1). We expect to get a wave that has both incoming and outgoing components:

$$u \sim a_{out} e^{i\omega r_*} + a_{in} e^{-i\omega r_*} \quad \text{as } r \to +\infty,$$ (4.19)

where a_{out} and a_{in} are functions of L and ω. It can be shown (problem 71) that these satisfy

$$1 + \left| a_{out}^2 \right| = \left| a_{in}^2 \right|.$$ (4.20)

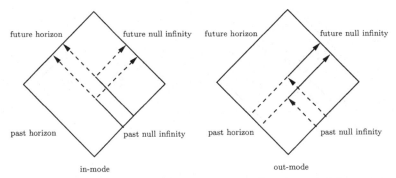

future horizon · future null infinity · future horizon · future null infinity

past horizon · past null infinity · past horizon · past null infinity

in-mode · out-mode

Figure 1   In-modes: rays from past infinity are scattered to future infinity and down the hole. Out-modes: incoming radiation from past infinity and the past horizon combine to ensure that there is no radiation crossing the event horizon (based on Frolov and Novikov, p. 93).

We can define transmission and reflection coefficients by

$$T = \frac{1}{a_{in}} \quad \text{and} \quad R = \frac{a_{out}}{a_{in}},$$

whence Eq. (4.20) becomes

$$|T|^2 + |R|^2 = 1. \tag{4.21}$$

Values for the transmission coefficient (corresponding to a wave propagating into the black hole) and the reflection coefficient (corresponding to radiation scattered to infinity) can be found by numerical calculation, or approximately in special cases. For very high frequencies ($\omega \gg 1/m$) the ω^2 term in the Regge-Wheeler equation (4.17) is much larger than the effective potential V_L in the vicinity of $r = 2m$potential term can be neglected and the waves are barely influenced by the hole. For very low frequencies ($\omega \ll 1/m$) the waves are almost entirely scattered back by the potential barrier in V_L. Figure 1 illustrates this solution. (Frolov and Novikov, page 93)

Other solutions corresponding to simple boundary conditions are illustrated in Fig. 2. The complex conjugates of the in-modes are the out-modes (corresponding to no radiation falling down the hole). We can define also 'up-modes' which are purely outgoing at infinity

$$u \sim e^{i\omega r_*} \quad \text{as } r_* \to +\infty,$$

and the 'down-modes' which are the complex conjugates of the up-modes.

We have dealt with test particles and test fields in a fixed metric background, ignoring the effect that such additional energy would have on the gravitational field. This approximation will not affect any of our conclusions below. But we mention in passing that consideration of the effect on the metric of a Schwarzschild black hole

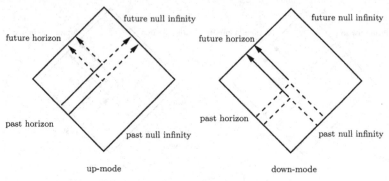

Figure 2   The up- and down- modes.

by a material perturbation leads to an equation of the form (4.17), although with a different effective potential (Zerelli 1970). The perturbed metric can be obtained from solutions of this equation. The results reveal a complex pattern of oscillations of the black hole referred to as quasi-normal modes. The details are beyond the scope of this book (Chandrasekhar 1983).

Of more immediate relevance to the topic of this chapter is the extension of the treatment of wave scattering to rotating black holes. On inserting the Kerr metric coefficients in to Eq. (4.13), the scalar wave equation becomes

$$\left[\frac{(r^2+a^2)^2}{\Delta}-a^2\sin^2\theta\right]\frac{\partial^2\psi}{\partial t^2}+\frac{4mar}{\Delta}\frac{\partial^2\psi}{\partial t\partial\phi}+\left[\frac{a^2}{\Delta}-\frac{1}{\sin^2\theta}\right]\frac{\partial^2\psi}{\partial\phi^2}$$
$$-\frac{\partial}{\partial r}\left(\Delta\frac{\partial\psi}{\partial r}\right)-\frac{1}{\sin\theta}\frac{\partial}{\partial\theta}\left(\sin\theta\frac{\partial\psi}{\partial\theta}\right)=0.$$

The equation can again be separated into normal modes as

$$\psi=R_{\omega LM}(r)S_{LM}(\theta)e^{iM\phi}e^{-i\omega t}.$$

We introduce an analogue of the coordinate r_*, in this case defined by

$$\frac{d}{dr_*}=\frac{\Delta}{r^2+a^2}\frac{d}{dr},$$

and a new radial variable

$$u(r)=(r^2+a^2)^{1/2}R_{\omega LM}(r),$$

where it is understood that $u(r)$ depends on the parameters ω, L and M, which we suppress for clarity. We obtain for $u(r)$ the radial equation

$$\frac{d^2u}{dr_*^2}+\{\frac{((r^2+a^2)\omega-aM)^2-\Delta(L(L+1)+a^2\omega^2-2aM\omega)}{(r^2+a^2)^2} \tag{4.22}$$
$$-\frac{r^2\Delta^2}{(r^2+a^2)^4}-\frac{d}{dr_*}\left(\frac{r\Delta}{(r^2+a^2)^2}\right)\}u=0. \tag{4.23}$$

This is the Teukolsky equation for a scalar wave; it reduces to the Regge-Wheeler equation when $a = 0$ (problem 70).

Now, we see from Eq. (4.23) that

$$u \sim e^{\pm i \omega r_*} \quad \text{as } r \to +\infty. \tag{4.24}$$

At the horizon ($r_* \to -\infty, r = r_+$) we have $\Delta = 0$, which implies that $r_+^2 + a^2 = 2mr_+$ and hence

$$\frac{d^2 u}{dr_*^2} + \left(\omega - \frac{aM}{2mr_+} \right)^2 u \sim 0.$$

Thus

$$u \sim e^{\pm i (\omega - M\omega_+) r_*} \quad \text{as } r \to r_+ \tag{4.25}$$

where $\omega_+ = a/(2mr_+)$ is the angular velocity of the hole, Eq. (3.29).

We can repeat the steps that led to Eq. (4.21). Now, however, we see that the transmitted wave (the one that goes down the hole, Eq. (4.25)) has a different wavenumber from the reflected wave (the component that is scattered to infinity, Eq. (4.24)). This is exactly analogous to the transmission and reflection of waves at the boundary between two different media: the transmission coefficient is multiplied by the ratio of wavenumbers such that

$$\frac{(\omega - M\omega_H)}{\omega} |T|^2 = 1 - |R|^2 \tag{4.26}$$

(problem 71). This leads to a novel feature that was not present in the spherical case. For wavemodes such that $\omega < M\omega_H$ the reflected amplitude exceeds that of the incidence wave (i.e. $R > 1$). Such waves are amplified by scattering, the energy coming from the rotation of the hole.

This amplification of waves by scattering is called superradiance. We can think of it as the induced emission of energy by the hole (induced by the incident radiation field). Recall that the induced emission of radiation from an excited atom in an electromagnetic field is a classical phenomenon. The effect of quantum mechanics is to cause spontaneous emission from the atom (emission independent of any external radiation field). We require both spontaneous and induced emission in order to achieve thermal equilibrium between an atom and a thermal radiation field. (Spontaneous and induced emission are described by the Einstein A and B coefficients, which must be related to each other to ensure thermal equilibrium with radiation having a Planck spectrum.) This prompts us to ask what might happen to a black hole in a thermal radiation bath. The surprise is not only that the inclusion of quantum theory does indeed lead to spontaneous emission from a rotating hole, but that, as we shall see later (section 4.5.4), a Schwarzschild black hole is also predicted to radiate.

For completeness note that equation (4.23) generalises from a spin zero (scalar) field to fields of arbitrary spin s, and that in the general case it is called the Teukolsky equation. For example, spin $s = 1$ corresponds to the electromagnetic field, spin

$s = 2$ to gravitational perturbations and spin $s = 1/2$ to (massless) neutrinos. It is a somewhat remarkable fact that this general wave equation in the Kerr background has separable solutions.

Problem 70 *Derive the asymptotic behaviour at large distances from the hole Eq. (4.24) and close to the horizon Eq. (4.25) of solutions of the scalar Teukolsky equation (4.23). Show that if we put $a = 0$ in the Teukolsky equation we recover the Regge-Wheeler equation.*

Problem 71 *For any two independent solutions of a second order linear differential equation, $u(r_*)$ and $v(r_*)$ (say) it can be shown that the quantity $W = u'v - v'u$ (where $'$ indicates d/dr_*) is constant. Evaluate W as $r_* \rightarrow -\infty$ by taking $u = u_{\omega L}(r_*)$ given by Eq. (4.18) and $v = u_{\omega L}^*$, the complex conjugate of u. Find W as $r_* \rightarrow +\infty$ by taking $u = u_{\omega L}(r_*)$ given by Eq. (4.19) and $v = u_{\omega L}^*$. Hence show that for the Regge-Wheeler equation (4.17) this leads to the relation (4.21) and for the Teukolsky equation (4.23) it leads to (4.26).*

4.4 Thermodynamics

The 'first law' of black hole mechanics Eq. (4.8) is rather reminiscent of the first law of thermodynamics: it says that the change in energy of a black hole δm comes partly from the change in rotational energy (the second term on the right) and partly from a term of the form $T\delta S$, provided that we identify (up to a constant of proportionality) the temperature T with the surface gravity κ and the entropy S with the surface area of the event horizon A_h. The constancy of κ on the horizon translates to the constancy of temperature between systems in equilibrium. And the area law, that the area of the horizon cannot decrease, is analogous to the second law of thermodynamics, that the entropy cannot decrease in a closed system. The analogy is consistent to the extent that this follows from the identification of the entropy of a black hole with its area through the first law. The problem with this analogy, in the context of classical general relativity, is that black holes are perfect absorbers but do not radiate: apparently, therefore, they cannot be assigned a 'real' temperature. This seemed to be an insuperable obstacle to making the formal analogy into a physical one.

The physical consistency of the picture is provided by the introduction of quantum mechanics. This raises the laws of black hole mechanics from mathematical analogies to physical descriptions through the Hawking process. According to this, black holes radiate like blackbodies at a temperature that depends on their mass (Eq. (4.29)).

In retrospect we can see why this should be so through various thought experiments, one of which we shall give below (following Kiefer, 1999). In effect these experiments make use of the fact that black holes can form parts of various physical systems and should not violate the thermodynamical behaviour of such systems. For example, if black holes have no entropy, it would be possible to reduce the entropy of a closed system containing a black hole by throwing things into the hole. In such circumstances, thermodynamics would not apply to systems containing black holes, for example, a star just past the point of collapse. (Incidentally, this situation is not rescued by the fact that an external observer never sees the final collapse: the entropy of a black hole, like its other attributes, relates to the properties of the spacetime on spacelike surfaces, (such as $U + V = $ const. in Fig. 6 of chapter 2), and not to what an external observer can see.)

There are numerous accounts of the Hawking process in the literature that attempt to explain what is happening using a simple quantum mechanical picture. These explanations are sometimes somewhat misleading. The Hawking radiation arises from a black hole that has been formed at a finite time in the past from the collapse of a star, and does not arise from an eternal black hole, as some of the simpler accounts might lead one to believe. Unless one arbitrarily imposes time-asymmetric boundary conditions, an eternal black hole is time-symmetric and therefore cannot radiate (since the emission of radiation would define a direction in time). Any picture that does not take the time-asymmetry of the collapse process into account cannot represent the quantum mechanics correctly. One well-known example of a dubious explanation is to imagine the quantum vacuum to be filled with short-lived particle-antiparticle pairs that come into existence for a time $\Delta E/h$ before annihilating again. In this time, one particle of negative energy falls into the hole and is captured while the other escapes to infinity as Hawking radiation. One can show that this picture leads to the correct order of magnitude for the black body temperature. However, any argument based on the uncertainty principle will lead to the correct functional form for the radiation on purely dimensional grounds. This does not prove the correctness of the physical picture. (For the sake of completeness we should add that the picture can be resurrected with the correct definition of particle and vacuum state, which are different for the collapsing star and for the eternal hole. The correlations in the eternal vacuum mean that the putative positive energy particle is cancelled out by interference with a negative energy particle from the past horizon, so there is no radiation. This takes us however beyond the scope of this book.)

In the next section we shall provide an argument that shows how the black hole temperature is required for consistency with thermodynamics. We shall then outline the quantum calculation which shows that black holes do have a temperature and hence that black holes do satisfy the laws of thermodynamics.

Note the parallel with Planck's original discovery of the quantum theory: here too classical physics was found to be inconsistent with thermodynamics, in this case with the thermodynamics of radiation, a situation that was resolved by the introduction of quantum theory.

There is perhaps one note of caution we should add. While we have stated the majority view, there is a minority opinion that this is not so much wrong as unproven. There are a variety of niggles, but the main one concerns the problem of trans-Planckian modes which we mention is section 4.5.4.

Problem 72 *Derive an expression for the Hawking temperature $T_h \propto \frac{hc^3}{GMk}$ by a dimensional argument, given that the temperature is inversely proportional to the black hole mass.*

4.4.1 Horizon temperature

We can show that it is necessary to associate a temperature with the horizon for consistency with thermodynamics (Keifer, 1999, Bekenstein 1980). A box containing a mass m of thermal radiation at a temperature T is lowered from a distance towards a spherically symmetric black hole. For simplicity we assume that the mass of the walls of the box and of the rope can be neglected. As we saw in section 2.3.8 (Eq. (2.24)) at a distance r from the black hole the energy in the box measured from infinity is given by

$$\mathcal{E}_r = m_0 \left(1 - \frac{2m}{r} \right)^{1/2}.$$

As the box approaches the horizon at $r = 2m$ the energy tends to zero, so the box gives up an energy m_0 in the process of being lowered from infinity.

Let the box now be opened and an amount of thermal radiation δm_0 be allowed to escape into the hole. This energy does not change the mass of the hole, as it is zero as seen from infinity. If the box is now raised, an energy of $m_0 - \delta m_0$ is expended to raise it back to infinity. So an energy δm_0 is dumped into the black hole, and at the end of the cycle δm_0 has been converted into external work at infinity. Therefore thermal radiation is being converted into work in a reversible cycle. The efficiency of the cycle is

$$\eta = \frac{\text{work out}}{\text{energy expended}} = \frac{\delta m_0}{\delta m_0} = 1,$$

which violates the second law of thermodynamics. Note that the sequence of operations does not violate the first law, but the violation of the second law would enable us to use the process to construct a perpetual motion machine of the second kind. In essence, disordered energy (radiation) is being converted to ordered work thereby decreasing the entropy of the Universe. Since an energy δm_0 is dropped into the hole arbitrarily close to the horizon the black hole mass is not increased, so its area does not increase and neither therefore does its entropy. But the entropy of the box is lowered when it returns to infinity. Thus the reversible cycle leads to a decrease in the entropy in the world.

The resolution turns out to be to question how arbitrarily close the box can approach the horizon. In fact, the box must have a finite size since it must be larger than the wavelength of the enclosed radiation. This limits the distance to which the box can approach the hole. As a result we shall see that only part of the energy δm_0 can be converted into work.

Radiation at temperature T has its peak intensity at a wavelength

$$\lambda \sim \frac{hc}{kT} \tag{4.27}$$

according to Wien's law. This is where quantum theory comes in (as is evident from the appearance of Planck's constant in the formula). The box must have a proper length of at least λ. Therefore it can be lowered to within a coordinate distance l of the horizon, where

$$\lambda = \int_{2m}^{2m+l} \frac{dr}{(1 - 2m/r)^{1/2}} = \int_{2m}^{2m+l} \frac{r^{1/2}dr}{(r - 2m)^{1/2}} \sim (2m)^{1/2} \int_0^l \frac{d\varepsilon}{\varepsilon^{1/2}} \sim 2\sqrt{2ml}.$$

Thus, rearranging,

$$l \sim \frac{\lambda^2}{8m}. \tag{4.28}$$

We now recalculate the energy of the box before and after opening it. Before opening it is

$$\mathcal{E}_r = m_0 \left[1 - \frac{2m}{(r + l)}\right]^{1/2} \sim m_0 \left[1 - \frac{2m}{(2m + l)}\right]^{1/2} \sim m_0 \left(\frac{l}{2m}\right)^{1/2}$$

$$\sim m_0 \frac{\lambda}{4m}$$

as $r \to 2m$, assuming $l \ll 2m$ and using Eq. (4.28). After opening the box the energy is

$$\mathcal{E}'_r \sim (m_0 - \delta m_0)\frac{\lambda}{4m}.$$

The efficiency η with which the thermal radiation can be transformed into work is $[\delta m_0 - (E_r - E'_r)]/\delta m_0$, or

$$\eta = \frac{\delta m_0 - \frac{\lambda}{4m}\delta m_0}{\delta m_0}$$

$$= 1 - \frac{\lambda}{4m}.$$

Using now the value for λ, Eq. (4.27), this becomes

$$\eta = 1 - \frac{hc}{4mkT},$$

which has the familiar form $\eta = 1 - T_s//T$ for a sink at temperature T_s if the black hole has a temperature

$$T_h = T_s \sim \frac{hc}{4mk}.$$

That the final relation is only an approximation comes from the fact that we have used only an order of magnitude argument. We can deduce however that the temperature of the black hole in physical units satisfies

$$T_h \propto \frac{hc^3}{GMk}.$$

For a Schwarzschild black hole $\kappa = 1/(4m) = c^4/(4GM)$, where κ is the surface gravity, so in this case $T_h \propto h\kappa/ck$. (In fact this expression holds in general, not just for a spherical hole.) So in order for a black hole to satisfy the laws of thermodynamics it would need to have a temperature given by the above expression.

4.4.2   The four laws of black hole thermodynamics

We are now in a position to state the laws of thermodynamics for black holes.

Zeroth law

The temperature of a black hole is constant on the horizon. This follows because $T_h \propto \kappa$ and the surface gravity κ is constant on the horizon (as we showed in section 4.1.1).

First law

The first law of black hole mechanics, Eq. (4.8) can now be written as

$$dm = T_h dS_h + \omega_+ dJ,$$

where $dS_h \propto dA_h$. We can therefore associate with a black hole an entropy S_h proportional to its surface area. Equivalently, $S_h \propto m^2$. With slightly more generality we can add an extra term for the charged black hole corresponding to the work done in acquiring a charge dq at an electrostatic potential Φ. The relation becomes

$$dm = T_h dS_h + \omega_+ dJ + \Phi dq.$$

This is the first law of black hole thermodynamics. Note that these arguments cannot determine the constants of proportionality in the expressions for temperature and entropy.

Second law

The direct translation of the area theorem in general relativity, that the area of a black hole cannot decrease, would be that the entropy of a black hole cannot decrease. This is not valid once quantum mechanics comes into play, because the Hawking radiation carries off mass and therefore reduces the surface area of the black hole and hence its entropy. The correct statement is a generalised form of the law, that the entropy of the universe, including that of the black hole S_h, cannot decrease i.e. if S_{ext} is the entropy of the world excluding the black hole then $S_h + S_{ext}$ cannot decrease. For example, for an isolated radiating black hole a detailed calculation shows that, although the entropy of the black hole decreases, the increase in entropy in the external world coming from the emitted radiation exceeds the loss in entropy of the hole, so this generalised form of the second law is obeyed (Bekenstein 1975).

Third law
The Planck statement of the third law of thermodynamics, that the entropy of any system tends to zero at zero temperature, is not valid, at least in classical black hole thermodynamics. It is violated by the extremal charged hole and the extremal Kerr hole (which have $T = 0$, $A_h \neq 0$). (The restriction to classical physics arises because extreme holes might not exist in whatever the quantum theory turns out to be.) In any case, the alternative Nernst version of the third law, that $T = 0$ cannot be reached in a finite number of steps, can be shown to hold for the temperature T_h of a black hole horizon. We have seen an example of this in our discussion of the spinning up of a black hole (section 3.16.1). Note for $a = m$ that $T_h = 0$ (since $\kappa = 0$). So the third law corresponds in this case to the impossibility of spinning up a black hole to $a = m$.

4.5 Hawking radiation

4.5.1 Introduction

In this section we outline a derivation of the properties of the radiation from a Schwarzschild black hole that has been formed by collapse. The result of this calculation is that the absolute values of the temperature and entropy can be determined. We find that a black hole radiates like a blackbody with a temperature in physical units given by

$$T_h = \frac{\hbar\kappa}{2\pi ck} \tag{4.29}$$

and that its entropy is given by

$$S_h = \frac{1}{4}\frac{kc^3}{\hbar G}A_h,$$

where, as usual, $\hbar = h/2\pi$.

The details of Hawking's original paper (Hawking 1974) can be followed with a basic knowledge of quantum field theory and complex analysis. The result has since been rederived in various ways and this reinforces the conclusion to a certain extent. On the other hand, the fact that no one derivation is utterly convincing has raised some doubts and there are certainly some open problems. However, the consistency of the result with thermodynamics means that almost no-one doubts the conclusions. Rather than repeat the details of one or more of the derivations in the literature, we shall try to give some insight into the main points of the physics while avoiding the full quantum field theory treatment.

The calculation treats the behaviour of an incoming quantum wave as it passes through a collapsing star and out to future (null) infinity. The major contribution to radiation from the hole comes from massless particles, since there is no threshold energy for the production of massless particle-antiparticle pairs. So, for example, for (hypothetical) massless scalar particles we should use the relativistic wave equation Eq. (4.12). Nevertheless, as long as we treat both particles and antiparticle waves explicitly, we can think of a non-relativistic many particle state (or quantum field) described by Schrödinger wave functions. Now, Schrödinger waves scatter off the gravitational field of the black hole just like classical waves, so this is not the essential ingredient introduced by quantum theory. Rather the key effect of quantum theory is felt through the properties of the quantum vacuum.

One way to understand how the vacuum has properties is to think in terms of zero point energy. This is usually introduced in the context of an isolated harmonic oscillator. Suppose however that the oscillator is coupled to an external system, for example that the oscillator consists of an electrically charged mass on a spring, which therefore interacts with the electromagnetic field. Even in its ground state the oscillating mass would radiate electromagnetic waves and so the external system would appear to drain the zero point energy from the oscillator. This would be inconsistent with quantum mechanics, in particular with the uncertainty principle. The resolution is that the electromagnetic field also has a zero point energy and just as the fluctuations in the oscillator radiate to the field, so zero point fluctuations in the field re-energise the oscillator, thereby maintaining overall balance. (This, incidentally, is why the ground state of an atom is stable.)

We can generalise this to the realisation that all fields have associated zero point energies. In fact, in relativity, we have not only the existence of positive energies but also of negative energies, or, equivalently, of positive and negative frequencies (since energy and frequency are related by a factor $\hbar$). We can see this from Eq. (4.14) where solutions $e^{\pm i\omega t} = e^{\pm iEt/\hbar}$ are permitted (and not only $e^{-iEt/\hbar}$ as for the non-relativistic Schrödinger equation). We can now describe the results of bona fide quantum field theoretical calculations by returning to Dirac's pre-field theory picture of the quantum vacuum. In Dirac's picture the vacuum consists of modes or states of positive and negative energy, with the negative energy states completely filled and states of positive energy empty. An anti-particle in this picture appears as the absence of a negative energy wave. (A little license is required here since Dirac's picture really applies only to fermions, so the description should not be taken too precisely.)

The crucial point will be that the quantum vacuum, in particular the negative frequency zero point modes, can be modified by the (classical) gravitational field of external matter.

4.5.2   Casimir effect

The best know and best established case of the distortion of the quantum vacuum is the Casimir effect. Here the electromagnetic field interior to a system of electrical conductors, two parallel plates for example, is modified by the necessity to satisfy conducting boundary conditions on the plates. Between the conductors only certain field modes (wavelengths or, equivalently, frequencies) can exists and each contributes its $\frac{1}{2}\hbar\omega$ to the zero point energy, whereas in the exterior region all modes are present. The difference in energy density is observed as a difference in pressure, which creates forces on the conductors (pushing together parallel plates or expanding a spherical conductor). The system is static so there is no radiation of energy. But at the microscopic level one would, in principle, find real fluctuating currents in the conductors, the presence of which is required for consistency with the zero point fluctuations in the fields.

4.5.3 Thermal vacua in accelerated frames

The usual vacuum state of a field in the absence of gravity is defined with respect to inertial observers in flat spacetime. Instead of cluttering up this vacuum with conductors to demonstrate the effects of negative energies, we can look instead at how it appears to an accelerating observer. As we shall see, this quantum system is similar in some respects to a black hole. It is therefore a useful preparation before we treat the effects of a real gravitational field. We confine ourselves to the simplest case of a uniformly accelerated observer in one space dimension and to the vacuum states of a scalar field (rather than the electromagnetic field). Even here we shall be content to outline the discussion, leaving the interested reader to find the details in the literature (Birrell and Davies 1982, Takagi, 1986, Unruh 1976).

The scalar wave equation (i.e. the relativistic Schrödinger equation) in one space dimension for the field Φ in the usual inertial (or Minkowski) coordinates (t, x) is

$$-\frac{\partial^2 \Phi}{\partial t^2} + \frac{\partial^2 \Phi}{\partial x^2} = 0.$$

The separable solutions will be of the form

$$\Phi(t, x) = u_\omega(t, x) \sim \{ \begin{array}{l} e^{\pm i\omega(x-t)} \\ e^{\pm i\omega(x+t)} \end{array},$$

with $\omega > 0$. (For comparison with the perhaps more familiar form $\exp \pm i(\omega t + kx)$ note that with $c = 1$ the wavenumber k satisfies $k = \pm \omega$.). Just as in non-relativisitic quantum mechanics we think of Φ as describing both a wave field and a particle state of definite momentum.

The vacuum for an inertial observer is defined by the absence of positive energy particles (or waves). However, the vacuum cannot be entirely devoid of energy, in some sense, because, even in the vacuum, fluctuations of the field must be present to maintain consistency with the uncertainty principle (corresponding exactly to the zero point energy of a harmonic oscillator). We can extend Dirac's original prescription (for fermions) and imagine that in the vacuum the positive energy states are empty but the negative energy states are populated. In fact, crudely speaking, we can think of the $\frac{1}{2}\hbar\omega$ of an oscillator to correspond to half a particle per frequency mode. In essence, we cancel this positive energy by populating the vacuum with half a negative energy particle per mode.

Positive energy solutions are eigenfunctions of the energy operator with positive eigenvalue, so

$$i\hbar \frac{\partial}{\partial t} u_\omega^{(+)} = \hbar \omega u_\omega^{(+)}$$

for positive energy solutions, and hence the time dependence of $u_\omega^{(+)}$ has the form

$$u_\omega^{(+)} \sim e^{-i\omega t},$$

with $\omega > 0$. So the allowed negative energy modes in the vacuum state are

$$\Phi \sim u_\omega^{(-)} = \frac{1}{2(\pi\omega)^{1/2}} e^{i\omega t \pm i\omega x}, \quad \omega > 0. \tag{4.30}$$

Note that we have not discussed normalisation of the wave function so we have not justified the various constants in Eq. (4.30), which are included for completeness. (The normalisation condition in relativity is not the same as for the non-relativistic Schrödinger equation. The reason is that we have to chose a normalisation that is preserved in time, and this depends on the evolution equation for the field. We shall ignore this technical detail here and below.) Note also that the wavenumber can be positive or negative since the negative energy waves can be incoming or outgoing. Equivalently, we can include both cases by writing

$$u_\omega^{(-)} = \frac{1}{2(\pi |\omega|)^{1/2}} e^{i|\omega| t - i\omega x}, \tag{4.31}$$

with $-\infty < \omega < +\infty$.

To find the eigenstates of the field for the uniformly accelerated observer, we transform the wave equation to the accelerated frame. This is achieved by using the Rindler coordinates (section 2.8.3). Since we are now using (t, x) for inertial coordinates we label the Rindler coordinates here as (η, ξ) instead of the (t, ξ) of section 2.8.3. We have

$$t = \xi \sinh a\eta$$
$$x = \xi \cosh a\eta \tag{4.32}$$

with

$$d\tau^2 = a^2\xi^2 d\eta^2 - d\xi^2.$$

Note that η is the proper time on the worldline $\xi = 1/a$ of the observer with proper acceleration a. The wave equation becomes

$$-\frac{1}{a^2\xi^2} \frac{\partial^2 \Phi}{\partial \eta^2} + \frac{1}{\xi} \frac{\partial \Phi}{\partial \xi} + \frac{\partial^2 \Phi}{\partial \xi^2} = 0, \tag{4.33}$$

with $c = 1$ as usual. This has separable solutions with positive energy with respect to the time η

$$U_{\omega'}^{(+)} = \frac{1}{2(\pi |\omega'|)^{1/2}} e^{-i|\omega'| \ln \eta} \xi^{i\omega'/a}, \qquad (4.34)$$

(with a normalisation appropriate to the relativistic wave equation) and corresponding negative energy solutions $U_{\omega'}^{(-)}$. These are the energy eigenstates according to the accelerated observer.

These solutions (4.34) form a complete orthonormal set at a fixed time η in the sense that

$$\int U_{\omega'}^{(+)} U_{\omega''}^{(+)*} \frac{d\xi}{\xi} \propto \delta(\omega' - \omega''),$$

where $\delta(x)$ is the Dirac delta function. Note the factor of ξ in the integrand is required for orthogonality.

We can prove this as follows:

$$
\begin{aligned}
\int U_{\omega'}^{(+)} U_{\omega''}^{(+)*} \frac{d\xi}{a\xi} &= \frac{1}{2(\pi |\omega'|)^{1/2}} \frac{1}{2(\pi |\omega''|)^{1/2}} e^{-i|\omega'|\eta + i|\omega''|\eta} \int \xi^{i\omega'/a - i\omega''/a} \frac{d\xi}{a\xi} \\
&= \frac{1}{2(\pi |\omega'|)^{1/2}} \frac{1}{2(\pi |\omega''|)^{1/2}} e^{-i|\omega'|\eta + i|\omega''|\eta} \int e^{i/a(\omega' - \omega'') \log \xi} d(\log \xi)/a \\
&= \frac{1}{2 |\omega'|^{1/2}} \frac{1}{|\omega''|^{1/2}} e^{-i\eta(\omega' - \omega'')} \delta(\omega' - \omega'') \\
&= \frac{1}{2\omega'} \delta(\omega' - \omega''),
\end{aligned}
$$

and we have used

$$\delta(x) = \frac{1}{2\pi} \int_{-\infty}^{\infty} e^{i\omega x} d\omega.$$

Now, an inertial negative energy wave of the form (4.31) can be expressed as a superposition of the complete set of states of positive energy waves $(U_{\omega'}^{(+)})$ and negative energy waves $(U_{\omega'}^{(-)})$ for the accelerated observer as

$$u_\omega^{(-)}(t(\eta, \xi), x(\eta, \xi)) = \sum_{\omega'} \{\beta_{\omega'}(\omega) U_{\omega'}^{(+)}(\eta, \xi) + \alpha_{\omega'}(\omega) U_{\omega'}^{(-)}(\eta, \xi)\}, \qquad (4.35)$$

where the $\beta_{\omega'}(\omega)$ and $\alpha_{\omega'}(\omega)$ are constants to be found. Strictly the sum in Eq. (4.35) is really an integral over continuous values of ω', but since we are interested in a mode by mode analysis we do not need to be concerned with such details.

We come now to the input from physics. Consider a quantum field which is in the mode $u_\omega^{(-)}$ of the inertial vacuum state. According to standard quantum theory, the coefficient $\beta_{\omega'}(\omega)$ on the right of (4.35) gives the amplitude for measuring a (positive frequency) particle (in the state $U_{\omega'}^{(+)}$) in the accelerated frame.

To obtain β we multiply through by $\xi^{-1} U_{\omega''}^{(+)*}$ and integrate with respect to ξ. Note that we include the factor ξ^{-1} because we want to use the orthogonality condition (4.34). The functions $U_\omega^{(-)}$ and $U_{\omega''}^{(+)*}$ can easily be shown to be orthogonal by repeating the calculation above (or simply replacing ω' by $-\omega$ and noting that $\delta(\omega' + \omega'') = 0$ if ω' and ω'' have the same sign). Thus, at any time η, for the accelerated observer in the inertial vacuum there is an amplitude for finding a particle of frequency ω' from each wave of frequency ω of

$$\beta_{\omega'}(\omega) = 2\omega' \int u_\omega^{(-)} U_{\omega'}^{(+)*} \frac{d\xi}{\xi} = \frac{1}{2\pi} \left(\frac{\omega'}{\omega}\right)^{1/2} \int e^{i|\omega'|\eta} \xi^{-i\omega'/a} e^{i|\omega|t - i\omega x} \frac{d\xi}{\xi}, \qquad (4.36)$$

where x and t are considered to be functions of ξ at time η through (4.32) and the integral is over the 'volume' of space.) Therefore according to the accelerated observer the inertial vacuum contains an overall number of particles of frequency ω' from all of the modes of the inertial vacuum

$$n(\omega') = \int |\beta_{\omega'}(\omega)|^2 \, d\omega. \qquad (4.37)$$

The integral for $\beta_\omega(\omega')$ can be evaluated in terms of gamma functions and hence the probability of detecting particles of a given frequency ω' found from Eq. (4.37).

However, we can simplify the calculation by making use of the fact that, since the Rindler metric is independent of time, the probability for the accelerated observer to detect particles must be independent of time. So we can evaluate this at a convenient time, which is $t = \tau = 0$. Thus we need

$$\beta_{\omega'}(\omega) = \frac{1}{2\pi} \left(\frac{\omega'}{\omega}\right)^{1/2} \int e^{-i\omega\xi} \xi^{-1 - i\omega'/a} d\xi, \qquad (4.38)$$

since $dx = d\xi$ on $t = 0$. To circumvent the detailed evaluation of the integral, note that, putting $z = i\omega\xi$, Eq. (4.38) can be written

$$\beta_{\omega'}(\omega) = \frac{1}{2\pi} \left(\frac{\omega'}{\omega}\right)^{1/2} \int e^{-z} (-iz)^{-1 - i\omega'/a} \omega^{-1 + i\omega'/a} i \, dz.$$

Writing $-i = e^{-i\pi/2}$ we get

$$\beta_{\omega'}(\omega) = \frac{1}{2\pi} \left(\frac{\omega'}{\omega}\right)^{1/2} e^{-\pi\omega'/2a} \int e^{-z} (z)^{-1 - i\omega'/a} \omega^{+i\omega'/a} dz$$

$$= \frac{1}{2\pi} \left(\frac{\omega'}{\omega}\right)^{1/2} e^{-\pi\omega'/2a} I,$$

where the last equation defines I. Note the $\beta \to 0$ as $a \to 0$, consistent with the fact that the inertial observer should see just the vacuum. Thus

$$|\beta_{\omega'}(\omega)|^2 = \frac{e^{-\pi\omega'/a}}{4\pi^2} \left(\frac{\omega'}{\omega}\right) |I|^2,$$

and

$$\left|\beta_{-\omega'}(\omega)\right|^2 = -\frac{e^{\pi\omega'/a}}{4\pi^2}\left(\frac{\omega'}{\omega}\right)|I|^2 = -e^{2\pi\omega'/a}\left|\beta_{\omega'}(\omega)\right|^2.$$

Hence,

$$n(-\omega') = -e^{2\pi\omega'/a}n(\omega').$$

A function satisfying this relation is

$$n(\omega') = \frac{1}{e^{2\pi\omega'/a} - 1} \tag{4.39}$$

which is a thermal spectrum. In one space dimension the factor for the volume of phase space in the Planck spectrum, usually $4\pi k^2 dk$, is just dk ($= d\omega$ with $c = 1$) The Planck distribution in one space dimension is therefore

$$n(\nu)d\nu = \frac{2\pi d\nu}{e^{h\nu/kT} - 1},$$

where $\omega' = 2\pi\nu$ (since ω' is in radians per unit time and ν is in cycles per unit time). Comparing this with Eq. (4.39) we conclude that the accelerated observer sees a thermal distribution of particles at temperature $\hbar a/(2\pi k)$, or

$$T_a = \hbar a/(2\pi kc),$$

reinstating physical units for the acceleration a.

Note that the calculation applies to an observer undergoing constant proper acceleration for all time in the inertial vacuum, and does not include any transient effects from starting and stopping the acceleration. The calculation does not tell us directly what an observer accelerated from rest can detect or what a system accelerated from rest will emit. Furthermore, while the discussion is instructive, it cannot be used directly to infer the properties of a black hole.

Problem 73 *Verify that (4.34) is a solution of the wave equation (4.33).*

4.5.4 Hawking radiation

In fact, one might think that it does not matter to a wave at a large distance from a spherical gravitating mass whether the mass is a star or a black hole. A static 'star' in the inertial vacuum cannot be emitting Hawking radiation (whatever an accelerated observer might see) by the very fact that it is static. The black hole is different: its interior is not static since the metric coefficients depend on time (recall that the r coordinate is timelike inside the horizon), or, equivalently, because it possesses an event horizon (which divides the static exterior from the non-static interior). One can see clearly from the spacetime diagrams (Figs. (7) and (10) of chapter 2) that the spacetime structure at infinity is not the same for both the star and the black hole. Physically, Hawking describes it by saying that rays traced backwards from late times pile up near $v = 2m + t_*$ (where t_* is the time at which the surface of the star crosses the horizon). This is roughly equivalent to saying that the late stages of collapse are visible for the whole of the future of an observer at infinity.

At past infinity, far from the collapsing star, we can define our usual inertial vacuum (the in-vacuum): this consists of negative energy states of the form

$$u_{\omega LM} = \frac{1}{2(\pi\omega)^{1/2}} e^{i\omega v} \text{ as } r_* \to \infty, \tag{4.40}$$

with $\omega > 0$. For an observer at future infinity, we shall see that the *presence of the horizon* means that the vacuum is a different state (the out-vacuum). As in the preceding section, we need to calculate how these negative energy states appear to such an observer.

The standard approach uses the Heisenberg picture in quantum mechanics, in which the states are fixed and the field operators are evolved backwards to work out the future content of the in-vacuum. Here instead we shall evolve the states as in the Schrödinger picture in introductory quantum mechanics. Rather than solve the wave equation however, we make use of the geometrical optics approximation to follow changes in the wave. This is possible because waves reaching infinity at late times are redshifted, so arise from waves that in the vicinity of the collapsing star have a high frequency, for which the geometrical optics approximation is valid. Also, spacetime near the horizon is flat (section 2.8.3) so the geometrical optics approximation is simplified.

Consider the behaviour of a negative frequency incoming wave $e^{i\omega v}$ as it travels through the star, to emerge just at the point where the outer shell crosses the horizon, and continues to future infinity. Waves beyond $v = v_0$ go down the hole, so only waves from $v < v_0$ contribute. For simplicity we can choose the labeling such that $v_0 = 0$.

At past (null) infinity the metric is asymptotically flat in Schwarzschild coordinates. Thus we measure the frequency of a wave by a uniform spacing in Schwarzschild time t or, equivalently, by the phase difference in $\omega v = \omega(t + r_*)$. At future infinity, near $r = 2m$ we have seen that the metric is also flat. In fact, we can see that in Kruskal coordinates the metric takes the Minkowski form near $r = 2m$, so we can use our usual notions of geometrical optics near the horizon, provided that we use these coordinates and not the Schwarzschild ones. Then the frequency of a wave will be measured by constant spacing in the Kruskal U coordinate, or, equivalently by the phase $\omega'U$. In geometrical optics the phase is constant along a ray in spacetime (by definition of rays!). Changes is frequency are measured by changes of spacing between the rays. This is illustrated in Fig. 3 for some ingoing and outgoing waves.

Now consider a high frequency wave that comes in from infinity near $v = 0$ passing through the star as it is about to cross the horizon and emerges highly redshifted to future infinity. Note that this is illustrated in the figure by reflection at the centre of the star. This is an artefact of a spacetime diagram showing a radial

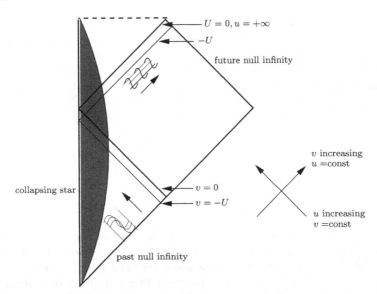

Figure 3 Penrose–Carter diagram of a star collapsing to a black hole showing a ray that passes through the star and grazes the horizon at $U = 0$.

coordinate and time. The incoming ray passes directly through the star, but, because the angular coordinates are suppressed, what we see in the diagram is the incoming ray at decreasing radial coordinate reflected into an outgoing ray at increasing radial coordinate.

We need to connect the form of the outgoing wave with that of the incoming wave. It is easy to see qualitatively what happens. The rays that fill the infinite future of an observer at a large distance from the hole are those that pile up at $v = 0$. Any ray that starts later than $v = 0$ goes straight down the hole and does not get to infinity. (We are ignoring any scattering here.) Thus a finite amount of v time in the distant past must be stretched to an infinite amount of u time in the far future. The figure shows quantitatively how this happens. The ray that grazes the horizon at a separation (in affine parameter) $-U$ arises from a ray that originates at early times far from the star at a parameter separation $v = -U$ from the last ray at $v = 0$. An incoming wave of the form $e^{\pm i\omega' v}$ (4.40) therefore becomes an outgoing wave of the form $e^{\pm i\omega' U}$.

We now express the outgoing wave $e^{i\omega' U}$, arising from the purely negative frequency incoming mode $e^{i\omega' v}$, in terms of the waves $e^{\pm i\omega u}$ that define the out-vacuum in the spacetime exterior to the hole. The two expressions are connected by the relation between Kruskal and Schwarzschild coordinates Eq. (2.54):

$$e^{i\omega' U} = e^{-i\omega' \exp(-\kappa u)},$$

where κ is the surface gravity.

We must therefore expand $e^{-i\omega' \exp(-\kappa u)}$ as a superposition of waves of the form $e^{\pm i\omega u}$ to find the positive and negative frequency content at future infinity of the purely negative frequency wave $e^{i\omega' v}$ at past infinity. The calculation follows that of section 4.5.3. We put

$$e^{-i\omega' \exp(-\kappa u)} = \sum_\omega \{\beta_\omega(\omega')e^{-i\omega u} + \alpha_\omega(\omega')e^{i\omega u}\},$$

where the coefficients $\alpha_\omega(\omega')$ and $\beta_\omega(\omega')$ are to be found. Taking the inverse Fourier transform, we obtain

$$\beta_\omega(\omega') \propto \int_{-\infty}^{+\infty} e^{i\omega u}e^{-i\omega' \exp(-\kappa u)}du, \tag{4.41}$$

which gives the amplitude for finding a positive frequency state $e^{-i\omega u}$ in the outgoing waves, starting from the incoming vacuum. Thus we can obtain the probability of finding particles at future infinity, in frequency (or momentum) state ω, by taking the modulus squared of the amplitude and integrating over all incoming frequencies. The result is proportional to

$$\int |\beta_\omega(\omega')|^2 \, d\omega'.$$

The number of particles emitted by the hole is also proportional to this probability.

To evaluate the integral for $\beta_\omega(\omega')$ in Eq.(4.41) put $x = e^{-\kappa u}$. We get

$$\beta_\omega(\omega') \propto \int_0^\infty x^{-1-i\omega'\kappa^{-1}}e^{-i\omega x}dx.$$

This is exactly the integral we studied in Eq. (4.38) for the case of the accelerating observer (with x here replacing ξ). Following the same procedure we therefore obtain a flux of particles (number per unit time in the frequency interval $d\omega$) with a spectrum proportional to

$$\frac{d\omega}{e^{2\pi\omega/\kappa} - 1}, \tag{4.42}$$

which is that of a black body with temperature proportional to κ.

Note that we have approximated the spacetime as flat at large distances and close to the horizon. More accurately we should use the Regge-Wheeler equation (4.17) or the Teukolsky equation (4.23) for the propagation of waves in the Schwarzschild or Kerr geometries. If the potentials in these equations are taken properly into account then there is an overall factor in Eq. (4.42) arising from the scattering of the waves off of the potentials. Since the potentials are different for particles of different spin the rate of emission of particle species from the black hole will differ, but the spectrum (i.e. the distribution of particles over different frequencies) will remain the same. These transmission factors are ignored below.

The outline of the calculation we have given here is clearly far from rigorous. The underlying calculation has now been picked over many times and redone in many different ways with some differing views as to what the conclusion should be. One of the outstanding problems appears to be the following. The radiation received by a distant observer can be traced back to close in to the black hole where it is exponentially blue shifted. Eventually, close enough in, the energy in the radiation should be not only large enough to distort spacetime, but large enough – beyond the Planck energy – to doubt the validity of classical gravity. The argument from quantum field theory seems to lead to these 'trans-Planckian' modes and hence to contain the seeds of its own undoing. These doubts however appear to be outweighed by the way that the inclusion of Hawking radiation leads to a consistent picture of thermodynamics. The expectation is that all doubt will be allayed once the quantum theory of gravity is discovered.

Problem 74 *Near $r = 2m$ show that*

$$\frac{r_*}{2m} - 1 \approx log_e \left(\frac{r}{2m} - 1 \right)$$

and that

$$1 - \frac{2m}{r} \approx \frac{r}{2m} - 1.$$

Deduce that

$$\kappa^2 \xi^2 \approx \exp \left(\frac{r_*}{2m} - 1 \right)$$

and hence the Kruskal coordinate $U = -e^{-u/4m}$ satisfies

$$U \propto \xi e^{-\kappa t}.$$

By comparing with the transformation from Rindler coordinates (t, ξ) to flat space (Minkowski) null coordinates $u = T - X$ (and $v = T + X$) deduce that the metric near $r = 2m$ has approximately the Minkowski form in Kruskal coordinates. Confirm this directly (and more simply!) from the Kruskal form of the metric near $r = 2m$.

4.6 Properties of radiating black holes

4.6.1 Entropy and temperature

The quantum mechanical calculation fixes the constants in the expression for the temperature and entropy of a black hole. Restricting ourselves to the Schwarzschild case, in physical units these are

$$T_h = \frac{\hbar c^3}{8\pi kGM} \sim \frac{10^{23}}{M} \text{ K}$$

or about $10^{-7}(M_\odot/M)$ K and

$$S_h = \frac{1}{4}\frac{kc^3}{\hbar G}A_h \sim 10^{77}\left(\frac{M}{M_\odot}\right)^2 k \quad \text{JK}^{-1}.$$

It is of interest to estimate the thermal entropy of a star before collapse to a black hole. If the star contains N particles the entropy is of order $S \sim Nk$. For a solar mass star $(M_\odot \approx 10^{30}$ kg$)$ this is $S \sim 10^{57}k$. The difference between the entropy of the star and the black hole it forms is therefore enormous. What is the source of this entropy? It is too large to be produced by dissipation during collapse. It is presumably associated with the gravitational microstates of the black hole either through the number of ways in which a black hole of given m and a can be formed - the number of initial states - or through the number of states within the horizon.

We can define 'natural' units of mass (or energy), length and time in terms of the fundamental constants G, $\hbar$ and c, respectively the Planck mass (or energy), length and time, as

$$m_P = \left(\frac{\hbar c}{G}\right)^{1/2}, \quad E_P = m_P c^2, \quad l_P = \left(\frac{\hbar G}{c^3}\right)^{1/2}, \quad t_P = \left(\frac{\hbar G}{c^5}\right)^{1/2}.$$

In terms of these quantities the horizon temperature is

$$\frac{kT_h}{m_p c^2} = \frac{1}{8\pi}\left(\frac{m_P}{M}\right)$$

and the entropy is

$$S_h/k = \frac{1}{4}\left(\frac{A_h}{l_P^2}\right).$$

Thus $S_h/k = \frac{1}{4} \times$ the area of the black hole in units of the square of the Planck length $(\hbar G/c^3)^{1/2}$.

Problem 75 *Show that the entropy of a Planck mass spherical black hole is $4\pi k$.*

4.6.2   Radiating black holes

The rate of energy loss by a Schwarzschild black hole is obtained from Stephan's law as follows:

$$L_h = A_h \sigma T^4 = 16\pi\left(\frac{GM}{c^2}\right)^2 \sigma \left(\frac{\hbar c^3}{8\pi GMk}\right)^4$$

where we have considered electromagnetic radiation only as this will make a major contribution to the emitted radiation. Inserting numerical values we get for the luminosity of the hole

$$L_h = 3.6 \times 10^{32}M^2$$

$$= 9.0 \times 10^{-29}\left(\frac{M_\odot}{M}\right)^{-2} \text{W}.$$

Since this must come from a decrease in the mass of the hole, we have

$$\frac{dM}{dt} = -\frac{3.6 \times 10^{32}}{c^2 M^2}.$$

integrating gives the lifetime of a black hole of mass M as

$$t_h \sim 8.33 \times 10^{19} \left(\frac{M_i}{10^{12} \text{ kg}}\right)^3 \text{ s}$$

where M_i is the initial mass of the hole. So a black hole of mass between 10^{11} kg and 10^{12} kg would be exploding today.

Problem 76 *What is the lifetime of a solar mass black hole?*

Problem 77 *Show that a black hole of mass 1.6×10^{12} kg has a lifetime of about 10^{10} years (a Hubble time). What is the lifetime of a black hole that is just hot enough to be emitting gamma rays?*

Problem 78 *What is the temperature of a Planck mass black hole? What is the entropy of a black hole of area 1 m^2.*

Problem 79 *The specific heat of a body of mass M is defined as $d(Mc^2)/dT$. Show that the specific heat of a Schwarzschild black hole of mass M is*

$$C = -\frac{8\pi k G}{\hbar c} M^2.$$

Note the negative sign: a black hole heats up as it radiates as is often the behaviour of gravitating systems. (The paradox is resolved by noting that we are leaving out of account the gravitational energy.)

4.6.3 Black hole in a box

Consider first a black hole in thermal equilibrium with a black body radiation field held at a fixed temperature, for example in a large thermal reservoir. The radiation will be affected by the gravitational field of the hole because corresponding modes of the field are relatively redshifted at different radii. The parameter T in the Planck distribution is therefore not the local temperature, but the temperature at infinity. The standard form of the Planck distribution takes this into account without modification.

Suppose that a fluctuation causes the emission of the black hole to increase by a small amount. Its mass will fall and its temperature will rise. But accompanying the fall in mass will be a decrease in the area of the black hole, so it will absorb less energy from the radiation field, the temperature of which we have supposed not to vary. Thus the departure from equilibrium grows. Likewise a fluctuation that causes the emission of the black hole to decrease by a small amount will shift the black hole away from equilibrium in the opposite sense. In this situation, therefore, thermal equilibrium is unstable. This outcome is a consequence of the negative specific heat of the black hole (Hawking 1976).

Nevertheless there are other circumstances under which a black hole can come into thermal equilibrium with radiation in a closed system. Consider a body radiating into an isolated enclosure initially containing no radiation. As the body radiates it heats up the enclosure. A normal hot body will cool as it radiates so the body and the enclosure will eventually come into thermal equilibrium at a common temperature. This follows no matter how large the enclosure: the larger the enclosure the lower the equilibrium temperature. On the other hand, the temperature of a black hole increases as it radiates into the enclosure. So can the temperature of the enclosure catch up that of the black hole to achieve a state of thermal equilibrium?

The answer is that thermal equilibrium can be achieved only if the volume of the enclosure is less than a certain maximum size. For larger volumes the black hole evaporates away before an equilibrium can be reached. The following demonstration is based on the treatment of Custodio and Horvath (2003).

For a black body of initial mass M_i in an enclosure of volume V containing blackbody radiation with energy density $\rho(t)$, conservation of energy implies

$$M_i c^2 = M(t)c^2 + \rho(t)V,$$

where $M(t)$ its mass of the black hole at time t. Thus, using $\rho = aT^4$, the temperature of the radiation at time t is given by

$$T(t) = \left[\frac{(M_i - M(t))c^2}{aV} \right]^{1/4}.$$

The temperature of the black hole at time t, from section 4.6.1, is $T = B/M(t)$ where $B = 1.12 \times 10^{22}$ kg K. So the condition for thermal equilibrium is

$$\left[\frac{(M_i - M_{eq})c^2}{aV} \right]^{1/4} = \frac{B}{M_{eq}},$$

which can be rearranged to give

$$\left(1 - \frac{M_{eq}}{M_i} \right) \left(\frac{M_{eq}}{M_i} \right)^4 = \frac{aVB^4}{c^2 M_i^5}. \tag{4.43}$$

The left hand side of this equation has a maximum value of 0.082. This imposes a maximum value on the right hand side and hence, for a given initial black hole mass, on the volume V of the box. For a box larger than the maximum size the black hole evaporates away before thermal equilibrium is reached.

Problem 80 *Confirm that the maximum value of the left hand side of Eq. (4.43) is 0.082. Deduce that the right hand side of Eq. (4.43) has a maximum value and hence that there is a maximum volume of a box in which a given mass black hole can reach equilibrium. (For a larger box the black hole evaporates before equilibrium can be reached.) What is the maximum volume of a box in which a Planck mass black hole can reach equilibrium?*

4.7 Entropy and microstates

In statistical physics entropy has a meaning in terms of the number of microstates corresponding to a given macrostate. One of the major outstanding problems of black hole physics is to find the way to count black hole microstates that will correspond to this view of entropy. That the problem is far from straightforward can be seen by observing that black holes are the unique physical systems for which the entropy is associated with an area rather than a volume. In the early days of black hole physics, following Hawking's original discovery of black hole radiation, there were high hopes that black holes would play for unified theories of gravity the role that radiation physics had played in Planck's discovery of the quantum theory. Indeed, the Hawking formula for the black hole temperature is the unique piece of physics involving gravity (G), relativity (c) and quantum theory (h). Recent calculations of the black hole entropy from string theory suggest that this hope might yet be born out. The details are beyond the scope of the present text, but we can (very) briefly outline the idea.

In string theory, the basic elementary objects are strings and certain membranes (or just 'p-branes' for membranes of dimension p) in high dimensional space-times. (So strings are 1-branes.) The higher dimensions are required for the consistency of the quantised theory and the branes carry the sources of the various fields, just as points (0-branes) carry the charges which are the sources of the electromagnetic field in our four dimensional spacetime. In fact, just as charge is quantised, so the source of the electric field is just a number (the number of 0-branes), so too do the higher dimensional sources come as numbers of various branes.

We make contact with the familiar four dimensional world of low energy physics by tightly rolling up the extra dimensions ('compactifying' them). When we cannot see the fine details (as in ordinary physics below the Planck energy) then combinations of strings and branes in the higher dimensions appear as an effective theory of curved spacetime and fields in four dimensions. A special role is played by the so-called D-branes on which strings can end (where 'D' stands for Dirichlet, not for a numerical value).

The extremal Reissner-Nordström black hole provides an important example. Like the extreme Kerr black hole, it has a Hawking temperature $T_h = 0$, so is non-radiating, hence static. From the point of view of its string theory formulation, it turns out to be a stable quantum object. In our low energy four dimensional world the spacetime contains an event horizon and an electric field. In the higher dimensional view it consists of D-branes and strings wrapped round a torus. In fact, to get the non-singular horizon of the Reissner-Nordström solution, it is necessary to have large numbers of branes and to wrap them round the compact dimensions many times. This is important: it is because this can be done in many ways for the *same* Reissner-Nordström black hole that gives the hole its internal degrees of freedom.

At low energies then, gravity is important and we get the Reissner-Nordström solution. At high energies gravity is relatively weak. We can get to this regime therefore by reducing the coupling between the objects of the theory and gravity to zero. In this limit the solution no longer describes a black hole horizon but is expressed instead in terms of certain D-branes moving in flat spacetime (because the non-gravitating D-branes do not generate curvature). The crucial point is that in passing from strong to weak coupling the density of states for a stable object such as an extreme Reissner-Nordström black hole remains unchanged. Since we are ignoring curvature the counting of different configurations becomes a matter of standard physics. In this way we get the volume of phase space Ω for the black hole and hence its entropy $S = k \log \Omega$. The result corresponds exactly to Hawking's entropy in various models, such as four and five dimensional extreme Reissner-Nordström black holes and even for those close to extreme. As of this writing the Schwarzschild black hole (and other non-extremal cases) however cannot be treated in this way.

Problem 81 *Show that the surface gravity of a Reissner- Nordstrom black hole is* $\kappa = \frac{mr_+ - q^2}{2r_+^3}$ *and hence that the temperature of an extreme Reissner-Nordstrom black hole is zero.*

Chapter 5

WORMHOLES AND TIME TRAVEL

5.1 Introduction

A spacetime metric, obtained from Einstein's equations, is a purely local quantity, which does not determine the global topology of spacetime. To illustrate the meaning of this statement take the analogy of a flat sheet of paper of zero thickness rolled into a cylinder. The metric on the flat sheet and on the cylinder are the same flat Euclidean metric, as is clear from the fact that distances between neighbouring points are not altered by rolling up the sheet without stretching or tearing it. But global relations *are* changed: points near opposite edges on the flat sheet, that were separated by the length of the sheet, are close together on the cylinder. The cylinder provides a short cut between certain points that were previously far apart. We can even (in principle) roll the cylinder up in its other dimension to form a torus, which is also locally indistinguishable from flat space as far as its metric properties are concerned. To avoid confusion, recall that a two dimensional surface may appear curved from the point of view of the three dimensional space in which it is embedded although measurements confined to the two-dimensional space, for example on the sum of the angles of a triangle, will show it to be intrinsically flat. Both the cylinder and torus are intrinsically flat although they appear to be curved from the point of view of the embedding space.

We can imagine rolling up (two-dimensional) Minkowski spacetime in a similar way. We can also generalise the construction to four dimensions. Consider two inertial observers at relative rest. We identify points along these worldlines at the same times so that the intervening space is wrapped up into a cylinder. A traveller arriving at event 1 in Fig. 1 finds himself at event 2, so there are now two paths between 1 and 2: the identification has provided the additional short cut between the two locations. We call this shortcut a wormhole. This is not a time machine or a means of faster than light travel. The outcome arises from the topology of a cylindrical spacetime: the distance between the two points the short way round is zero.

5.2 Wormholes

The problem with the model of the previous section is that we do not have a practical means of identifying points in Minkowski spacetime to create the initial shortcut. Can

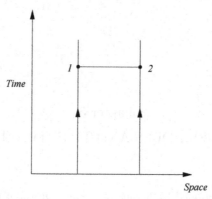

Figure 1   Points at the same time on the two world-lines are identified, that is, deemed to be the same event, thereby creating a shortcut in zero time between the world-lines.

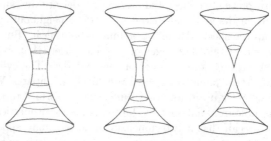

Figure 2   The interior of the Schwarzschild solution is dynamic: this is represented in the embedding diagram as a pinching off of the throat.

wormhole metrics be created from gravitating matter and made to behave in the same way?

Our first thought might be to take the complete Schwarzschild spacetime (defined by the Kruskal metric) and roll it up such that the region IV in Fig. 6 of chapter 2 is identified as part of our own universe and not a separate one. We could then see if we could travel through the black hole to a different region of spacetime, possibly in such a way that enables us to arrive at a distant destination faster than a light ray taking the conventional route. This proves to be impossible. Figure 2 shows why. The embedding diagram of Fig. 9 in chapter 2 does not take into account the change in the Schwarzschild metric with time interior to the event horizon. When we attempt to cross from region I to region IV, it turns out that the throat pinches off just as we arrive to go through it. Another way of saying this is that however we travel through the hole there is no way to avoid the singularity.

The Kerr solution (Fig. 6 of chapter 2) seems to offer additional possibilities to avoid the singularity, which is here timelike (section 3.18). But in this case the space-time structure is unstable to the small perturbations which would arise if we were to try to move any gravitating matter or energy between asymptotic regions. The instability alters the spacetime structure and would appear to lead to a spacelike singularity (Frolov and Novikov, 1998).

Our conclusion is that black holes will not enable us to construct short cuts across the Universe. However, it turns out that there are other ways in which we might achieve this through what are called traversible wormholes.

5.3 Traversible wormholes

In this section we look at metrics for traversible wormholes. We shall approach the issue just as in previous chapters by giving the metric of a simple example of a wormhole spacetime and investigating its properties. However, in contrast to our approach in chapters 2 and 3, we shall not insist that the space-times are empty of matter, since this would rule out any interesting wormhole solutions. We shall postulate a metric and use Einstein's equations to tell us the properties of the gravitating material needed to produce it, in contrast to the usual procedure by which the metric is obtained by solving Einstein's equations for a given matter distribution.

A general metric for a wormhole takes the form

$$d\tau^2 = e^{2\Phi(r)}dt^2 - \frac{dr^2}{1 - b(r)/r} - r^2\left(d\theta^2 + \sin^2\theta d\phi^2\right) \tag{5.1}$$

where Φ and b are functions of r. The metric is spherically symmetric and static. The minimum value of r is $r_{\min} = b_0$ which satisfies $r = b(r)$. This defines the throat of the wormhole. At infinity $\Phi(r) = 0$; otherwise it is negative and finite, so a stationary particle always has a timelike worldline: there is no horizon. The proper distance from the throat is l given by

$$dl = \frac{dr}{(1 - b(r)/r)^{\frac{1}{2}}}. \tag{5.2}$$

A simple special case of a wormhole metric is obtained by putting $\Phi = 0$, $b = b_0^2/r$, with $r^2 = l^2 + b_0^2$,

$$d\tau^2 = dt^2 - dl^2 - \left(b_0^2 + l^2\right)\left(d\theta^2 + \sin^2\theta d\phi^2\right) \tag{5.3}$$

which is the case we shall discuss. The embedding diagram for this metric is given in Fig. 3. In contrast to the Schwarzschild metric, which is time-dependent inside the horizon, the wormhole metric is static everywhere. Thus the embedding diagram gives the properties at all times. An observer can travel through the throat and emerge at the other side in a finite time (problem 83). There is therefore no singularity in the way. This is why the spacetime is referred to as a traversible wormhole. An alternative way to demonstrate the absence of a singularity is to calculate the curvature. This also turns out to be finite everywhere.

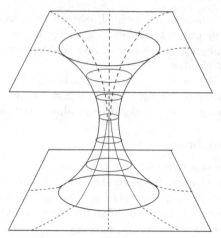

Figure 3   The embedding diagram for the special wormhole metric (5.3).

Notice that for outwardly propagating radial light rays $dt = dl$. Thus a radial light ray can always escape to infinity, so this is another way of seeing that the spacetime has no horizon. There are two regions, $l \to \pm\infty$, where the curvature tends to zero at large distances from the wormhole. These regions are joined by the throat of the wormhole. There is no restriction on travel between the two regions.

How could this provide a short cut across space? To see in principle how it could be done recall that we are free to change the topology without affecting the metric. Figure 4 shows how we can (in theory) join spacetime to create a shorter route through a wormhole.

Problem 82 *Show that the embedding diagram of Fig. 3 is given by $z = b_0 \cosh^{-1}$ (r/b_0). Calculate the circumference of the throat of the wormhole.*

Problem 83 *In the wormhole metric (5.1) show that relative to a stationary observer the velocity of a body with energy per unit mass E in radial free fall is given by*

$$v = \left(1 - \frac{e^{2\Phi}}{E^2}\right)^{\frac{1}{2}}. \tag{5.4}$$

(This illustrates that as long as Φ is finite then there is no region where $v = c$ so there is no horizon.) Show that the coordinate time to traverse a wormhole from $-L$ to $+L$ is

$$\int_{-L}^{+L} \frac{dl}{e^{\Phi} v}. \tag{5.5}$$

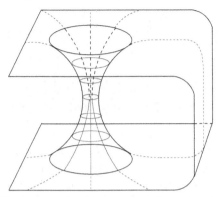

Figure 4   If the mouths of a wormhole connect regions of the same spacetime the wormhole provides a short-cut.

Problem 84 *Show that the radial coordinate of a light ray of energy at infinity E_{ph} and angular momentum L_{ph} in the simple wormhole metric (5.3) satisfies*

$$\left(\frac{dr}{d\lambda}\right)^2 = \left[E_{ph}^2 - \frac{L_{ph}^2}{r^2}\right]\left(1 - \frac{b_0^2}{r^2}\right).$$

By looking at the zeros of the right hand side, show that light with impact parameter $L/E = b_0$ falls into the wormhole, and hence that the capture cross-section for a light beam is πb_0^2.

Problem 85 *Show that the proper acceleration a felt by an observer stationary in the wormhole metric (5.1) at radial coordinate r is given by*

$$a = \pm\left(1 - \frac{b}{r}\right)^{\frac{1}{2}} \Phi'. \tag{5.6}$$

(Hint: Obtain $a^\mu = Du^\mu/d\tau$ and use $a^\mu a_\mu = -a^2$ (section 1.3.9).)

This shows that if $\Phi \neq$ constant, rocket motors are required to remain at rest; otherwise the acceleration is zero.

5.4   Creating a wormhole

Can the wormhole solutions of section 5.3 be created from gravitating matter and constructed to provide shortcuts across spacetime? A problem arises when we ask what sort of matter we need in order to create the curvature of the wormhole metric. Substituting the metric coefficients from (5.3) into Einstein's equations (1.14) shows us the matter content corresponding to the wormhole metric. This is quite a lengthy calculation the result of which is (Morris and Thorne, 1988)

$$T^{00} = T^{11} = -\frac{1}{8\pi G}\frac{b_0^2}{(b_0^2 + l^2)^2} \tag{5.7}$$

with the tangential components of equal magnitude but opposite sign. The quantity T^{00} is the density of mass-energy and T^{11} is the radial pressure. Note that the energy density and radial pressure are both negative!

Negative energy densities in the local rest frame of the matter are not possible in classical physics. More generally, classical physics forbids energy densities which are positive in the local rest frame but negative to rapidly moving observers. This is called the 'weak energy condition' which we deal with in the next section. In quantum mechanics negative energy densities are possible and we shall look at these in section 5.6.

5.5   Weak energy condition

Let v^μ be the 4-velocity of an observer and $T^{\mu\nu}$ the energy-momentum tensor of a fluid. Then the scalar product $T^{\mu\nu}v_\mu v_\nu$ is equal to T^{00} in the frame of the observer. If the observer is stationary with respect to the fluid then

$$T^{\mu\nu}v_\mu v_\nu = T^{00} = \rho, \tag{5.8}$$

where ρ is the mass-energy density of the fluid in its rest frame. If the observer is moving with respect to the fluid with velocity v in the $x-$direction then $v^\mu = (\gamma, \gamma v, 0, 0)$, $v_\mu = (\gamma, -\gamma v, 0, 0)$ and

$$T^{\mu\nu}v_\mu v_\nu = T^{0'0'} \tag{5.9}$$
$$= \gamma^2 T^{00} + \gamma^2 v^2 T^{11}. \tag{5.10}$$

Thus $T^{0'0'} = \gamma^2(\rho + v^2 p)$ where p is the pressure in the $x-$direction in the rest frame of the fluid.

The weak energy condition requires that $T^{0'0'} \geq 0$ i.e.

$$\rho + p \geq 0. \tag{5.11}$$

This ensures that all observers (all values of v) will measure a positive energy density, a condition that 'normal' matter would be expected to obey. Conversely, in the wormhole spacetimes the weak energy condition is violated, so some observers will measure a negative energy density. We call material with a negative energy density 'exotic matter'. The possibility of creating wormholes therefore comes down to the viability of creating stable exotic matter.

Problem 86 *The stress-energy tensor of a perfect fluid is given by*

$$T^{\mu\nu} = (\rho + p)\, u^\mu u^\nu - p g^{\mu\nu}.$$ (5.12)

Verify that

$$T^{\mu\nu} v_\mu v_\nu = \gamma^2(\rho + v^2 p)$$ (5.13)

where v^μ is the 4-velocity of an arbitrary observer moving with speed v in the x-direction with respect to the rest frame of the fluid and $\gamma = (1 - v^2)^{-1/2}$.

Problem 87 *Vacuum energy has the equation of state $p = -\rho$, with ρ positive. Show that vacuum energy does not violate the weak energy condition. Show also that the vacuum energy has the same value in all inertial frames, as it should to be consistent with the principle of relativity.*

5.6   Exotic matter

Consider two ordinary flat metallic plates held parallel and separated by some distance a. These form a capacitor. There exists an attractive force between the plates which increases with decreasing separation. One can think of this as a force between the fluctuating currents in the metallic boundaries. These do not cancel because they are correlated, the currents in one plate inducing those in the other and vice versa. Since the force is attractive the energy density between the plates must be less than that of the exterior vacuum, that is, it must be negative:

$$\rho = -\frac{\pi^2 \hbar}{720 a^4}.$$ (5.14)

This is called the Casimir energy and has been extensively verified experimentally (Bordag et al. 2001). The corresponding energy density of the electromagnetic field on the horizon of a black hole is (Jensen et al. 1988)

$$\rho = -\frac{19 \hbar}{7680 \pi^2 M^4}.$$ (5.15)

The amount of negative energy involved in a laboratory-scale capacitor is minuscule compared to that required for a macroscopic wormhole. It is not clear if large amounts of negative energy, sufficient to form a macroscopic wormhole, can be created. However, there is at present no known reason why, in principle, this should not be possible. We shall assume in what follows that wormholes can be made by a sufficiently advanced technology. This assumption forces us to face a fundamental issue: if a wormhole can be constructed then so too can a time machine.

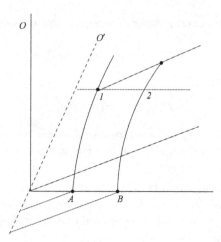

Figure 5 The figure shows the world lines A and B of the two wormhole mouths in their initial rest frame O. Also shown dashed are the space and time axes of the frame O' moving with speed v with respect to the frame O in the positive x-direction. The frame O' is the final rest frame of the mouths A and B. Note that the events 1 and 2 at which the acceleration ceases are simultaneous in O but not in O'.

5.7   Time machines

Let us assume that the topology of flat spacetime has been altered to contain a wormhole. In this section we explain how such a wormhole can be turned into a time machine. (Visser, 1995)

Consider two worldlines A and B at relative rest. These will approximate the two mouths of a traversible wormhole. As in section 5.1 to make the wormhole, we identify points on A and B that correspond to equal times in their rest frame O. The short cut through the wormhole takes zero time in this approximation: a traveller at an event on A has the choice of travelling to B through the external space or, taking a shortcut in zero time, through the wormhole. We now give each mouth of the wormhole the same constant acceleration at a time $t = 0$ in the initial rest frame O (Fig. 5).

Viewed from the frame O the acceleration comes to an end at some time t and the wormhole mouths have each acquired a speed v, say. Viewed from the new rest frame of the wormhole mouths O', mouth B comes to rest first at time $t'_b = \gamma(t - vx_b)$ followed by mouth A at the later time $t'_a = \gamma(t - vx_a)$ where we have used a Lorentz time transformation between frames. This is an example of the relativity of simultaneity. Therefore in O' the wormhole connects different times with mouth A ahead of mouth B by an amount $T = v\gamma(x_b - x_a) = \gamma v s$ where s is the constant separation of the two mouths in O; their distance apart in frame O' is γs.

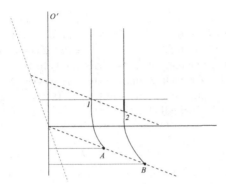

Figure 6   As for Fig. 5 but viewed from the final rest frame O'. The axes of O are dashed here. The wormhole now joins an event on A to an earlier event on B. This is represented in the inset.

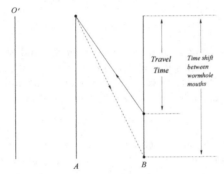

Figure 7   The figure shows a closed-timelike path along which an observer can travel into their own past.

However, the time through the wormhole is unchanged: it is still zero. We therefore have the situation depicted in Fig. 6.

To complete our construction of a time machine we must arrange matters so that the time shift is greater than the light travel time between A and B. To do this we imagine bringing the mouths closer together slowly enough that the time shift is not significantly altered. Figure 7 shows the state of affairs if the time lapse as a result of this construction is larger than the speed of light journey the long way round. It is now possible to leave B on a timelike path, arrive at A, and return via the shortcut wormhole route to a time before setting off.

A similar construction can be carried out in principle with 'real' wormholes in curved spacetime. To move the wormholes physically now, we could charge a wormhole mouth electrically, or shift it using gravity. Kinematically, observers stationed at the throats would experience the same changes to their spacetime under acceleration as did the Minkowski observers of our previous example, since the spacetime outside the throats is approximately flat. Notice that in the frame of an observer at one of the throats, the distance through the wormhole remains constant, so the travel time through the wormhole is unchanged. This corresponds to the picture of Fig. 4.

One can ask whether such a time machine is serviceable in the sense that it remains stable when normal matter moves into it. In fact, as long as the accelerations are sufficiently small, only a small change in the exotic matter content is required to keep the wormhole mouth open (Morris et al, 1988).

An alternative way of achieving large timelike separations between the mouths of a wormhole is to take one mouth of the wormhole on a round trip journey at high speed (Thorne, 1994). This employs the well-known twin effect whereby one twin moves into the future of the other twin. Suppose the time separation achieved by this means is T. When the travelling wormhole mouth arrives back, it is then possible to step into it and come out a time T in the past (less the time it takes to travel through the wormhole, which can be relatively short).

5.8 Chronology protection

Time machines raise a number of problems that go beyond mere paradoxes. They all concern the well-known classic problems of interfering with the past, for example preventing one's own birth. The creation of wormholes leads to the possibility of time machines which seem to imply logical inconsistencies. Where in the argument have we gone wrong?

There are several possibilities: (i) Wormholes that would allow time travel to the past are unstable: as soon as such a wormhole is created light can travel through it an infinite number of times contributing a positive energy density that collapses the wormhole. (ii) Wormholes are possible but the laws of physics (somehow) forbid inconsistent actions (such as interfering with one's own past). This is called the chronology protection conjecture. (iii) Wormholes are forbidden by as yet unknown laws of physics. This is a popular get-out but is easier said than believed, because it is difficult to see where such laws might intervene in the discussion.

Chapter 6
ASTROPHYSICAL BLACK HOLES

6.1 Introduction

In the preceding chapters we have explored the properties of two highly symmetric vacuum solutions of the field equations of Einstein's general theory of relativity, the Schwarzschild and Kerr black holes. For many years it was expected that these black holes would turn out to be artifacts of high symmetry and that real stars, being neither exactly spherically nor axially symmetric, would not collapse to black holes. In that case black holes would be artificial mathematical constructs. On the other hand, if real non-symmetric stars do collapse to black holes one might expect there to be lots of differently non-symmetric black holes for them to collapse to. But we already know that, as far as electrically neutral vacuum solutions are concerned, there are only the two black hole solutions, a property that, as we noted, is described by saying that black holes have no hair. The situation seems somewhat surprising.

The resolution of the paradox is even more surprising! It is that in relativity, in contrast to Newtonian gravity, a collapsing body grows increasingly symmetric as the collapse proceeds. The difference arises because in general relativity an asymmetric body radiates away energy in the form of gravitational waves (wave-like oscillations of the metric coefficients $g_{\mu\nu}$) and in so doing radiates away its asymmetry. The problem can be treated by looking at the behaviour of small non-axially symmetric perturbations of the gravitational field during the collapse process (Price, 1972). The result is that multipole moments of the matter distribution are radiated away leaving, in general in the exterior region, a Kerr black hole. More generally one can use topological arguments to show that a black hole must form, and hence that it must be a Kerr or Schwarzschild black hole (Hawking and Ellis, 1973).

According to general relativity therefore, black holes are described exactly by the Kerr metric (including the Schwarzschild metric as a special case). We do not expect charged black holes to exist in nature, as such an object would rapidly attract opposite charges and be neutralised. So there are two questions we need to answer: do black holes exist, and if so are they really Kerr black holes? In this chapter we review the observational evidence.

The earliest evidence for the existence of black holes has come from the study of binary star systems. Within the Galaxy a number of binary systems are known that consist of a visible star and a compact dark companion in close orbit about each other. In some of these systems the orbital parameters of the two bodies can be determined and a lower limit to the mass of the dark companion obtained. When this mass is found to be greater than the limiting mass for a stable neutron star (2-3 $M_\odot$) then the dark companion becomes a strong candidate to be a black hole: we know of no physical mechanism that can balance the inward pull of gravity in a compact 'dead' star once its mass exceeds $3M_\odot$. Several dozen such candidates have been found. However we can be certain that black holes really do exist only when we have demonstrated the existence of an event horizon, since this is the unique feature that defines a black hole and sets it apart from all other astronomical objects.

The discovery of stellar mass black hole candidates does not exhaust the possibilities. The study of active galaxies and also of the central region of our own Galaxy provides strong evidence for the existence of black holes with masses in the range $10^6 - 10^{10}\ M_\odot$, the so called supermassive black holes. It is now believed that such a massive black hole resides at the centre of nearly every galaxy including our own.

More recent evidence suggests that there may also be black holes with masses intermediate between stellar mass black holes and supermassive black holes. These objects have been detected within neighbouring galaxies and in two cases at the centres of globular clusters, the dense star systems that orbit about the Milky Way.

One other possibility is that the Universe could contain a population of low mass black holes, the so-called mini black holes. The evidence for these is so far inconclusive.

Overall, we shall see that the accumulated evidence for the existence of black holes is now very compelling.

Theories of gravity other than general relativity also predict the existence of black holes. The question whether the black hole candidates are Kerr black holes therefore constitutes a test of general relativity. This is a more difficult question to answer but, as we shall see, the x-ray emission of black hole binary systems does appear to accord with accretion by a Kerr black hole.

6.2   Stellar mass black holes

6.2.1   Formation

The theory of stellar evolution predicts that a star with a main-sequence mass exceeding about eight solar masses will end its life as a supernova, leaving behind a neutron star. In the cores of such massive stars nuclear burning goes all the way up to ^{56}Fe or ^{56}Ni (depending on the n:p ratio). These nuclei lie at the maximum of the nuclear binding energy curve, so energy-releasing reactions cease once this stage is reached. Once the mass of the iron core of such a star has built up to $1.4M_\odot$, the Chandrasekhar mass, the core will undergo gravitational collapse to form a neutron star. The gravitational energy released by the collapse appears initially as kinetic energy of the in-falling matter. Once nuclear densities are reached the collapse is halted, by nuclear forces and neutron degeneracy pressure, and the subsequent rebound launches a shock wave outwards through the core and into the envelope of the star. The shock wave heats the core to temperatures of the order of 10^{11} K but retains enough energy to blow off the envelope of the star. The result is a supernova. The bulk of the energy generated by the collapse is rapidly radiated from the very hot core in a large pulse of neutrinos that may also play a role in blowing off the envelope of the star.

In 1987 two neutrino detectors, one in Japan and one in the USA, simultaneously registered a short burst of about 12 neutrinos. A few days later supernova SN 1987A was observed in the large Magallenic cloud. The time delay and the numbers of neutrinos detected were consistent with the source of the neutrinos being SN 1987A at a distance of 50 kpc.

Although the equation of state of neutron matter is currently unknown, we can say that the upper limit to the mass of a neutron star cannot be greater than about $3M_\odot$. This estimate comes from setting the speed of sound in neutron matter equal to the speed of light; this condition leads to the stiffest (least compressible) equation of state consistent with causality. In fact the upper limit to the mass of a neutron star may not exceed $2M_\odot$ (Cottam et al, 2002).

It seems likely that the scenario outlined above for the formation of a neutron star and the accompanying supernova applies to the collapse of a star with a main sequence mass that is less than about $30M_\odot$. It now appears that the collapse of a more massive star forms a black hole rather than a neutron star. The signature of such an event is an exceptionally energetic supernova, or hypernova as it has been dubbed, accompanied by a short very intense gamma-ray burst. (Hjorth, 2003). These gamma ray bursts were first detected more than 30 years ago and have for some time been the subject of intense study. It is only recently that this connection between the long duration gamma-ray bursts, those lasting more than a few seconds, and supernovae has been established.

Another likely process for the formation of a black hole is the merger of the two members of a double-neutron-star binary system. The best known example of such a system is the binary pulsar discovered by Hulse and Taylor in 1974 (Taylor, 1994). Observations of this system have established that the two stars are slowly spiralling in towards each other. The system is losing energy and angular momentum, in the form of gravitational radiation, at a rate that is in agreement with the predictions of general relativity. It is estimated that the two stars will merge in about 3.2×10^8 years from now. As the total mass of the system is $2.83M_\odot$ the final product of the merger depends on the equation of state of neutron matter, which is currently unknown. If the equation of state is soft, that is if the material is more easily compressed, the result will be a black hole.

It is believed that the short duration gamma ray bursts, those having a mean duration of about 0.2 seconds, may be the result of such neutron star mergers. It is anticipated that the burst of gravitational radiation emitted in the final stages of a merger will be measurable with gravitational wave detectors such as VIRGO in Italy, LIGO in the USA and GEO in Germany.

Recently another neutron star binary pulsar has been discovered (Burgay et al, 2003) and found to contain two pulsars. This system has an orbital period of only 2.4 hours, a third of that of the Hulse-Taylor system, and an estimated time before the two stars merge of 8.5×10^7 years. The discovery of this new system has increased the expected rate of mergers. The detection rate by existing gravity wave detectors could be as high as once every 1-2 years.

6.2.2 Finding stellar mass black holes

An isolated black hole is completely dark so it can be detected only through its gravity. One way that this can be done is through the gravitational lensing of a background star when a dark object passes in front of it. The effect of a gravitational lens is to increase the intensity of light coming from the background star. The intensity increases smoothly to a maximum and then dies back down giving rise to a symmetrical light curve. If the distance to the lens can be obtained its mass can be calculated from the light curve. If the mass of the lens is greater than $3M_\odot$ then the object is a black hole candidate. Gravitational lensing has been observed, but to date none of the lens stars appears to be a black hole candidate.

All the known stellar mass black hole candidates are located in x-ray emitting binary star systems that consist of a star and a compact companion object in a close orbit about each other. In close binaries tidal forces tend to circularise the orbits. If the orbital parameters of the binary system can be obtained, a lower limit to the mass of the compact object can be estimated as follows. The time variation of the Doppler shift of absorption lines in the stellar spectrum yields the orbital period P of the system and also the amplitude of the line of sight velocity V_s of the star. These quantities determine the mass function

$$F(M) \equiv \frac{PV_s^3}{2\pi G} = \frac{\sin^3 \phi M_d}{(1 + M_s/M_d)^2},$$

where M_s is the mass of the star, M_d is the mass of the dark companion and ϕ is the angle between the line of sight and the normal to the orbital plane. Notice that $F(M)$ is a lower limit to M_d so if $F(M) > 3M_\odot$ we have a black hole candidate.

The first such black hole candidate to be discovered by this method was Cygnus X-1. It was discovered in 1965 and found to lie at a distance of 2.5 kpc from us. Its orbital period is 5.6 days. A number of independent estimates of the mass of Cygnus X-1 have been made using the mass function together with the assumed mass of the visible companion. They indicate a value somewhere in the range 6-20 $M_\odot$. The large uncertainty arises from the unknown inclination of the orbit to the line of sight. Nevertheless, this mass range makes Cygnus X-1 a strong black hole candidate.

Since the discovery of Cygnus X-1 dozens of other black hole candidates have been found in binary systems in our Galaxy. Recently the first example of an eclipsing

X-ray binary system, has been discovered in the nearby galaxy M 33 (Orosz, J.A. et al, 2007). The importance of this discovery is that the angle ϕ between the line of sight and the normal to the orbital plane is known to an accuracy of 1.5%. The mass of the companion is determined to be $70 \pm 7 M_\odot$ giving a mass for the dark companion of $15.65 \pm 1.45 M_\odot$ .This mass is well above the maximum mass for a neutron star making this object the strongest candidate to be a black hole yet discovered.

Now $F(M) \gtrsim 3 M_\odot$ is a necessary condition for an object to be a black hole, but on its own it is not sufficient. Definitive proof requires evidence for an event horizon since this distinguishes a black hole from an object with a surface. A systematic comparison of neutron star binaries with candidate black hole binaries has found a difference between the two types of binary that is ascribable to the presence of a solid surface in the neutron star case and its absence in the black hole candidate case.

In both types of binary the two stars are close enough together for material from the visible star to be pulled off and captured by the companion. Because the material has angular momentum it tends to orbit about the compact object forming a pancake shaped accretion disc. A packet of gas in the disc at radius r will move in a nearly Keplerian orbit with velocity $v \propto r^{-1/2}$. As the gas making up the disc has turbulent viscosity neighbouring annuli rubbing against each other will give rise to energy dissipation. This causes the gas to spiral slowly inwards to smaller radii and heats the gas in the inner part of the disc to temperatures high enough for it to emit x-rays. If the compact object is a neutron star the gas will eventually hit its surface and accumulate there. The kinetic energy of the in-falling gas is converted to heat as it strikes the surface of the neutron star and gravity on the stellar surface compresses the gas to the point where it undergoes thermonuclear burning. If the burning is unstable the result is a violent explosive outburst of x-ray emission. Such bursts are observed from neutron star binaries. We would not expect to see such an outburst from an accreting black hole as it has no surface. And no instance of such a burst has been seen from a black hole candidate (McClintock et al, 2003). This is evidence that the candidate black holes have an event horizon.

Problem 88 *Derive the expression quoted above for the mass function of a binary star system.*

6.2.3 The black hole at the centre of the Galaxy

Lynden-Bell and Rees suggested in 1971 that there might be a massive black hole at the centre of the Galaxy. In this section we look at the current evidence for this.

In 1974 a strong compact radio source was detected in the direction of the Galactic centre and the discovery was confirmed by further observations the following year. This source was named Sagittarius A* in 1982 by Brown.

The central few parsecs of the galaxy also contain a cluster of stars having a number density of about 10^6 pc^{-3}. The proper motions of individual stars in this central region have been followed over a period of several years and show that the stars are moving about the position of Sagittarius A*. Thus Sagittarius A* is located at the dynamic centre of the Galaxy. In particular two thirds of the orbit of a star called SO-2 has been plotted out. The orbit is highly elliptical and at its closest approach to Sagittarius A* is only 17 light hours from it. Its semi-major axis is 5.5 light days and the orbital period is about 15 y. The mass at Sagittarius A* needed to keep SO-2 in its Keplerian orbit comes out to be $(4.1 \pm 0.6) \times 10^6 M_\odot$. (Ghez et al, 2008). A central cluster of neutron stars having this mass within a radius of less than 17 light hours is ruled out by this result as the density of stars would have to be of the order of 10^{15} stars pc^{-3}. The stars of such a dense cluster would collide and merge to form a black hole within a short time scale. Other more exotic alternatives have been invoked such as a neutrino ball. However, as there is no independent evidence for the existence of these hypothetical objects it is hard to take them seriously. The conclusion is therefore that Sagittarius A* is a black hole.

We can go further than this. There is now some evidence that Sag A* may have a significant angular momentum. Recent observations of Sag A* in infrared radiation have detected flares which exhibit variability with a period of about 17 minutes (Genzel 2003). The flares also have rapid rise and decay times, which suggests that they are generated in the innermost zone of the accretion disc. If this is the case then the 17 minute period could be the orbital period of matter close to the last stable orbit of an accretion disc around the black hole.

As we can see from Eq. (3.50) in section 3.13.2 the orbital period depends on the spin of the hole mainly through its effect on the radius of the last stable orbit. For a Schwarzschild black hole of mass $3.7 \times 10^6 M_\odot$ the period of matter in the last stable orbit is about 28 minutes. So a period of 17 minutes is too short for a slowly spinning black hole of the mass of Sag A*. As we can see from Fig. 2 in chapter 3, the radial coordinate of the last stable orbit of a particle in a prograde orbit decreases with increasing angular momentum parameter a. In fact a 17 minute period is obtained for a prograde orbit when the black hole has $a = 0.52m$ (problem 46). Note from Fig. 2 that the radial coordinate of retrograde orbits increases with increasing a so a 17 minute period rules out retrograde orbits in this case.

The calculations of the orbital periods quoted above assume that the orbits lie in the equatorial plane. This is justified for the inner region of an accretion disc, since Lens-Thirring precession and viscous torques will force an out of plane orbit into the equatorial plane (Bardeen and Petersen, 1975).

Problem 89 *Verify that the period, measured by a distant observer, of a particle in the last stable orbit in the equatorial plane of a Schwarzschild black hole of mass* $3.7 \times 10^6 M_\odot$ *is 28 minutes.*

6.3   Supermassive black holes in other galaxies

Observations of the centres of other galaxies cannot be as detailed as those of the Milky Way but again it is found that material near their centres is orbiting rapidly

around a central mass concentration. In addition some galaxies emit large amounts
of energy from a very small volume at their centres. Quasars are an extreme example
of these active galaxies. The luminosity from the centre of a quasar can exceed the
output from the rest of the galaxy a thousand fold, yet this energy is coming from a
volume not much larger than that of the Solar System. The only mechanism that can
reasonably explain such a high luminosity from such a small volume is the accretion
of matter by a black hole at a rate of up to several solar masses per year. Recall,
from chapter 3, that a lump of matter that is accreted from the last stable orbit of
an extreme Kerr black hole must have lost about 30 per cent of its rest energy in the
process of spiralling in. On the other hand nuclear burning in stars can release no
more than 0.7 per cent of the rest energy of a star. It is hard to envisage how enough
stars to produce the observed luminosity could exist within such a small volume as
a stable system. On the other hand, a black hole explains the luminosities and other
features of active galaxies convincingly.

The Chandra X-ray observatory satellite has found two super-massive black
holes in orbit about each other in the galaxy NGC 6240. The exceptionally high rate
of star formation in this galaxy suggests that it formed recently from the merger of
two smaller galaxies. If each contained a central black hole this would explain the
presence of two black holes near the centre of the merged galaxy.

The process by which super-massive black holes form at the centres of galaxies
is not yet understood, but at the time of writing there are some clues (see section
6.3.1). They are thought to form at an early stage in the history of a galaxy and to
power the quasar activity observed in young galaxies (Barkana and Loeb, 2003).

A further intriguing possibility that is close to realisation is, in a sense, actually
to 'see' the event horizon of Sagittarius A*. The resolution of very long baseline radio
interferometers (VLBI) is approaching the scale of the event horizon of Sagittarius
A* (Doeleman et al, 2008). Calculations have been carried out that simulate the
appearance of the emitting gas surrounding Sagittarius A* (Melia and Falcke, 2001).
These show that under high resolution VLBI a central shadow of diameter $\sim 10m$,
for a black hole of mass m, would be visible in the radio image of the Galactic centre.
The detection of such a shadow would constitute direct evidence for the existence of
an event horizon.

6.3.1 Intermediate mass black holes

Over the past few years a number of highly luminous compact x-ray sources have been
detected in nearby galaxies. These sources emit 10-100 times as much x-ray power as
stellar mass black holes in x-ray binaries. At present their nature is unclear but they
may be accreting black holes with masses of the order of 100-$1000M_\odot$. In particular
they are often found in galaxies where an episode of rapid star formation is underway
- so called star-burst galaxies. This suggests that they may be formed by the merging
of stellar mass black holes produced in large numbers in the star burst. It has been

suggested that intermediate mass black holes could sink to the core of a galaxy where they would merge to form a supermassive black hole. These intermediate mass black holes could therefore be the key to understanding how the super-massive black holes form.

There is also evidence for black holes at the centres of globular clusters. Measurements with the Hubble space telescope of the velocities of stars in the central regions of two globular clusters point to the presence of black holes there. The putative black hole at the centre of the globular cluster M15, which is in orbit about our Galaxy, is estimated to have a mass of about 4000 $M_\odot$ (Van Der Marel et al, 2002). The very large globular cluster G1 attached to the Andromeda galaxy appears to harbour a central black hole of about $2 \times 10^4 M_\odot$ (Gebhardt et al, 2002). In both of these cases the relationship between the black hole mass and the cluster mass bears a similar relationship to that between the mass of a galactic black hole and the mass of its host galaxy.

6.3.2   Mini black holes

There is no lower limit to the mass of a black hole so in principle black holes with masses $<< 1M_\odot$ can exist. However the lower the mass of the black hole the greater the density to which matter has to be compressed to create it. This can be seen from Eq. (1.1) which states that $\rho \propto M^{-2}$. Take for example a black hole having a mass of 10^{12}kg. Its density must be of the order of 10^{57} kg m^{-3} (10^{39} times the density of a neutron star). Only in the very early universe, when ultra-high densities prevailed, could suitable conditions have existed for the formation of such small black holes.

From chapter 4 we recall that black holes emit Hawking radiation and have a lifetime $t \propto M^3$. Thus, any primordial black holes that exist today must have had an initial mass of at least 10^{12} kg, since any lighter black holes would have evaporated away before the present time.

How would an evaporating black hole nearing its final stage manifest itself? As its mass decreases its temperature increases according to the Hawking relation Eq. (4.29)

$$T_h = \frac{\hbar c^3}{16\pi kGM}.$$

The criterion for emission is that as soon as the temperature exceeds the rest mass of a particle species that species starts to be emitted by the black hole. At lower temperatures photons, neutrino antineutrino pairs and gravitons would be emitted. As the mass of the hole declines and the temperature increases then to these particles would be added successively electron-positron pairs followed by the heavier lepton anti-lepton pairs and then quark anti-quark pairs and beyond these ,if they exist, the supersymmetric particles. At some point the temperature reaches the stage where quantum gravity effects come into play. As we do not yet have a theory of quantum gravity the final end point of the evaporation process is not known. The end point could be a stable object with a mass close to the Planck mass, in which case such relics would contribute to the dark matter that makes up most of the mass of a galaxy.

What is the observational evidence for the existence of primordial black holes? The burst of gamma rays and other particles emitted in the final stage of evaporation of a single black hole are unlikely to be detected, but the gamma radiation from the population of such objects would contribute to the cosmic gamma ray background. Measurements of the gamma ray background give an upper limit to the present mass density of primordial black holes which is less than $10^{-8}\Omega$, where Ω is the cosmological density parameter having a value close to 1.

On the other hand the quarks emitted in the evaporation process would give rise to jets of hadrons that decay ultimately to photons, leptons, protons and antiprotons. These jets are similar to those produced in collisions of high-energy electrons and positrons at, for example, the LEP particle accelerator. The antiprotons could account for the unexpectedly large fraction of antiprotons that are detected in cosmic rays. Another anomaly is that the electron-positron cosmic ray background has a 50 per cent positron content in the vicinity of 100 MeV. Such a high ratio is hard to explain without a contribution from primordial black holes. (MacGibbon and Carr, 1991).

However, we must conclude that at present the evidence for the existence of primordial black holes is inconclusive.

6.4   Further evidence for black hole spin

In section 6.2.3 we cited some evidence for spin in Sagittarius A*. Here we look at some evidence for the spin of stellar mass black holes in binary systems.

We saw in chapter 3 that the angular momentum of a black hole strongly affects the orbits of particles and light rays close to the event horizon. Thus, for example, the spin of a black hole will influence the way in which matter from an accretion disc is captured. For an extreme Kerr black hole, in which the gas is orbiting in the same sense as the hole, the binding energy released by the gas as it spirals inwards towards the last stable orbit is about 30 per cent of its rest energy. On the other hand, for gas with angular momentum in the opposite sense to that of an extreme Kerr black hole the last stable orbit is at $r_{ls} = 9m$, and the binding energy liberated up to this point is only about 3 per cent. For a non-rotating black hole $r_{ls} = 6m$ and the binding energy released is about 6 per cent.

Many black hole candidates in binary systems emit an ultra-soft x-ray component with a black body (colour) temperature in the range 0.5 to 2 KeV. Zhang and Cui (1998) have suggested that the strength of this soft x-ray component is directly related to the spin of the black hole. They base their conclusion on the calculation of the x-ray flux from the surface of a thin accretion disc in the equatorial plane of a Kerr black hole. According to their calculations the temperature and luminosity of the soft x-rays are greatest for an extreme Kerr black hole rotating in the same sense as the accretion disc, prograde rotation, and least for the extreme retrograde case. Based on their soft x-ray spectra, black hole candidates can be assigned to three categories: extreme prograde, extreme retrograde and non-spinning or slowly spinning systems. However two of the black hole candidates, one of which is notably Cygnus X-1, appear to switch periodically from a retrograde system to a prograde system. Numerical simulations show that the switching of accretion flow is not as improbable as it might at first seem. Unfortunately the physics of accretion discs is not yet understood well enough for accurate models of the accretion process to be constructed.

Jets of ionised plasma travelling at relativistic velocities are observed from both stellar mass and super-massive black hole systems. The idea is that a pair of back-to-back jets are directed along the spin axis, or symmetry axis, of the black hole. Often only a single jet is visible. Since the jets are travelling at relativistic velocities, the idea is that aberration will enhance the luminosity of a jet with a component of motion in our direction, and will diminish the luminosity of the other jet. It is thought that the power source for jet production is the rotational energy of the black hole. We saw in chapter 3 that orbits having negative angular momentum and negative energy exist within the ergosphere region of a spinning black hole, and that these orbits can be used to extract energy and angular momentum from a black hole using the Penrose process. The Penrose process is however not a practical mechanism for energy extraction (section 3.15.4). It seems most likely that some other process powers astrophysical jets, probably involving the coupling of the rotation of the hole to exterior matter by magnetic fields as in the Blandford-Znajek mechanism. (For a non-technical account see Begelman and Rees, 1998.)

Current work on the measurement of spin involves the observation of lines of ionised iron in the x-ray region in the emission spectrum of active galactic nuclei. These lines are formed in gas close enough to the black hole for their spectroscopic profiles to be affected by general relativistic redshifts in addition to the normal Doppler shifts that arise from the motion of the gas. In principle the black hole parameters (mass and angular momentum) and the orientation of the accretion disc can be determined from these line profiles. In practice, the profiles depend on models of the emission region, and there may also be other contributions to the line broadening from scattered radiation. In addition the observations have limited resolution. So this work has some way to go.

In future x-ray polarisation measurements of the line emission from matter close to the black hole will also yield the parameters of the hole and act as a test of relativity.

6.5   Conclusions

The existence of stellar mass black holes, and of super-massive black holes at the centres of most, if not all, galaxies is supported by an abundance of evidence. There is little doubt that such objects exist as described by the general theory of relativity. Only recently has it been possible to detect intermediate mass black holes, and the evidence for them is slighter but growing. It is likely that these three categories of black holes are related. The current picture is that in the centres of galaxies, and of globular clusters, stellar mass black holes merge and grow to supermassive black holes and intermediate mass black holes respectively. On the other hand mini black holes are not the products of stellar evolution. Their existence would depend on the suitability of conditions in the early Universe for their formation, so their existence is of great interest for cosmology.

Solutions to Problems

Problem 1 The Galilean transformation for infinitesimals is $dx = dx' + vdt'$, $dy = dy'$, $dz = dz'$ and $dt = dt'$. Substituting this into equation (1.7) $ds^2 = c^2dt^2 - dx^2 - dy^2 - dz^2$ gives $ds^2 = c^2dt'^2 - dx'^2 - dy'^2 - dz'^2 - 2vdx'dt' - v^2dt'^2 \neq ds'^2$. A similar result follows for the sum of squares. In general the reason why we cannot find an invariant under the Galilean transformation is that the time component is fixed, so changes in the x component cannot cancel changes in the time component.

Problem 2 Multiplying $v'^\mu = \frac{\partial x'^\mu}{\partial x^\nu} v^\nu$ by $\frac{\partial x^\rho}{\partial x'^\mu}$ gives $\frac{\partial x^\rho}{\partial x'^\mu} v'^\mu = \frac{\partial x^\rho}{\partial x'^\mu} \frac{\partial x'^\mu}{\partial x^\nu} v^\nu = \delta^\rho_\nu v^\nu = v^\rho$. Change the free index ρ to μ and the dummy index μ to ν and the result follows. A similar procedure yields the inverse transformation of the covariant components.

Problem 3 The scalar product $v^\mu v_\mu = \frac{\partial x^\mu}{\partial x'^\nu} v'^\nu \frac{\partial x'^\rho}{\partial x^\mu} v'_\rho = \delta^\rho_\nu v'^\nu v'_\rho = v'^\nu v'_\nu$. Hence the scalar product is an invariant.

Problem 4

$$\nabla_\nu(A^\mu B_\mu) = \partial_\nu(A^\mu B_\mu).$$

Differentiating gives

$$B_\mu \nabla_\nu A^\mu + A_\mu \nabla_\nu B^\mu = B_\mu \frac{\partial A^\mu}{\partial x^\nu} + A^\mu \frac{\partial B_\mu}{\partial x^\nu}.$$

Now

$$\nabla_\nu A^\mu = \frac{\partial A^\mu}{\partial x^\nu} + \Gamma^\mu_{\nu\rho} A^\rho.$$

So

$$A^\mu \nabla_\nu B_\mu = A^\mu \frac{\partial B_\mu}{\partial x^\nu} - B_\mu \Gamma^\mu_{\nu\rho} A^\rho.$$

Interchange the indices μ and ρ in the last term and cancelling the A^μ gives

$$\nabla_\nu B_\mu = \frac{\partial B_\mu}{\partial x^\nu} - \Gamma^\rho_{\nu\mu} B_\rho.$$

Problem 5 Let

$$L_V A^\mu = V^\nu \nabla_\nu A^\mu - A^\nu \nabla_\nu V^\mu$$
$$= V^\nu \left[\frac{\partial A^\mu}{\partial x^\nu} + \Gamma^\mu_{\nu\rho} A^\rho \right] - A^\nu \left[\frac{\partial V^\mu}{\partial x^\nu} + \Gamma^\mu_{\nu\rho} V^\rho \right]$$
$$= V^\nu \partial_\nu A^\mu - A^\nu \partial_\nu V^\mu + \Gamma^\mu_{\nu\rho} A^\rho V^\nu - \Gamma^\mu_{\nu\rho} A^\nu V^\rho.$$

The last two terms cancel, as the dummy indices can be relabelled to make them identical, so it is legitimate to replace the partial derivatives by covariant derivatives in the Lie derivative.

Problem 6 The tensor transformation law for a second rank contravariant tensor is

$$T^{\mu\nu} = \frac{\partial x^\mu}{\partial x'^\sigma} \frac{\partial x^\nu}{\partial x'^\rho} T'^{\sigma\rho}.$$

The co-variant derivative is

$$\nabla_\sigma T^{\mu\nu} = \frac{\partial T^{\mu\nu}}{\partial x^\sigma} + \Gamma^\mu_{\sigma\rho} T^{\rho\nu} + \Gamma^\nu_{\sigma\lambda} T^{\rho\lambda}.$$

The Lie derivative is

$$\pounds_V T^{\mu\nu} = V^\sigma \nabla_\sigma T^{\mu\nu} - T^{\sigma\nu} \nabla_\sigma V^\mu - T^{\mu\sigma} \nabla_\sigma V^\nu.$$

Problem 7 We have $g^{\alpha\beta} g_{\alpha\beta} = 4$. Therefore

$$g^{\alpha\beta} \left(R_{\alpha\beta} - \frac{1}{2} g_{\alpha\beta} R \right) = R - 2R = -R = g^{\alpha\beta} T_{\alpha\beta} = 0. \tag{1}$$

Problem 8 We have from problem 6

$$\pounds_K g^{\mu\nu} = k^\alpha \nabla_\alpha g^{\mu\nu} - g^{\alpha\nu} \nabla_\alpha k^\mu - g^{\mu\alpha} \nabla_\alpha k^\nu$$
$$= 0 - (\nabla^\nu k^\mu + \nabla^\mu k^\nu),$$

from which the result follows.

Problem 9 From the metric the proper circumference of the curve r = constant in the plane $\theta = \pi/2$ at constant time t is $2\pi r$. At the event horizon $r = 2m$. So the proper circumference of a Schwarzschild black hole is $4\pi m$. From the metric the area of the surface r = constant, t = constant is

$$\text{Area} = r^2 \int_0^{2\pi} d\phi \int_{-\pi/2}^{\pi/2} \sin\theta d\theta.$$

The angular integral is the area of the unit sphere i.e. 4π and $r = 2m$ so the black hole area is

$$A_h = 16\pi m^2.$$

Problem 10 Expanding the scalar product $g^{\mu\nu}u_\mu u_\nu = 1$ gives

$$g^{00}(u_0)^2 + g^{11}(u_1)^2 + g^{33}(u_3)^2 = 1.$$

Inserting the values of the metric coefficients, putting $u_0 = E$ and $u_3 = -L$ and rearranging gives

$$\left(1 - \frac{2m}{r}\right)(u_1)^2 = \frac{E^2}{\left(1 - \frac{2m}{r}\right)} - \frac{L^2}{r^2} - 1.$$

Now $u_1 = g_{11}u^1 = -(1 - 2m/r)^{-1}dr/d\tau$ so finally we get

$$\left(\frac{dr}{d\tau}\right)^2 = E^2 - \left(1 + \frac{L^2}{r^2}\right)\left(1 - \frac{2m}{r}\right).$$

Problem 11 The geodesic equations are

$$\frac{du^\mu}{d\tau} + \Gamma^\mu_{\sigma\rho}u^\sigma u^\rho = 0.$$

The $\mu = 1$ equation, in physical units, is

$$\frac{d^2r}{d\tau^2} + \Gamma^1_{00}c^2\left(\frac{dt}{d\tau}\right)^2 + \Gamma^1_{22}\left(\frac{d\theta}{d\tau}\right)^2 + \Gamma^1_{33}c^2\left(\frac{d\phi}{d\tau}\right)^2 = 0.$$

With the Newtonian metric, $\Gamma^1_{00} = m/r^2$, $\Gamma^1_{22} = -r$ and $\Gamma^1_{33} = -r\sin^2\theta$ and at low speeds, $dt \sim d\tau$. With these values the $\mu = 1$ equation becomes

$$\frac{d^2r}{dt^2} = \frac{GM}{r^2} + r\left(\frac{d\phi}{dt}\right)^2$$

where the spherical symmetry confines the particle orbits to a plane and we have chosen the $\theta = \pi/2$ plane. This is the Newtonian expression for radial acceleration under gravity in polar coordinates.

Similarly the $\mu = 3$ equation becomes $r\frac{d^2\phi}{dt^2} + 2\frac{dr}{dt}\frac{d\phi}{dt} = 0$. This is the Newtonian expression for transverse acceleration in polar coordinates.

Problem 12 Expanding $1 = g_{\mu\nu}u^\mu u^\nu$ gives for radial motion

$$1 = g_{00}(u^0)^2 + g_{11}(u^1)^2.$$

Substituting for the metric coefficients with $u^0 = E/(1 - 2m/r)$ gives

$$1 = \left(1 - \frac{2m}{r}\right)\frac{E^2}{(1 - 2m/r)^2} - \frac{(u^1)^2}{(1 - 2m/r)}.$$

Finally, rearranging and taking square roots gives

$$u^1 = \pm\left[E^2 - \left(1 - \frac{2m}{r}\right)\right]^{1/2}.$$

Problem 13 The 4-velocity of an observer hovering at radius r is

$$u_H^\mu = \left(\left(1 - \frac{2m}{r} \right)^{-\frac{1}{2}}, 0, 0, 0 \right).$$

The covariant components of the 4-velocity of a particle in radial free fall are

$$(u_\mu) = (E, u_1, 0, 0).$$

So the energy per unit mass of the falling particle in the frame of the observer is given by

$$u_H^\mu u_\mu = \gamma = (1 - v^2)^{-\frac{1}{2}}.$$

where v is the local velocity of the particle. Evaluating the scalar product in the Schwarzschild frame gives $E \left(1 - 2m/r \right)^{-\frac{1}{2}}$. But the scalar product is an invariant, so we can equate it to its value γ in a local inertial frame, hence

$$E \left(1 - 2m/r \right)^{-\frac{1}{2}} = \left(1 - v^2 \right)^{-\frac{1}{2}}.$$

The velocity at r of a particle falling from infinity is just the velocity needed to project the particle back from r to infinity i.e. the escape velocity; hence setting $E = 1$, the energy of a particle dropped from rest at infinity, gives

$$v_{esc} = \left(\frac{2m}{r} \right)^{\frac{1}{2}}.$$

At the horizon the escape velocity is the speed of light.

Problem 14 Doing this analytically is rather lengthy so an easier way is to plot the effective potential against r (see figure 2) and note that the maximum occurs at the smaller value of r.

Problem 15 (a) Invert equation (2.26) to get L in terms of r and the result follows. (b) Substitute for L in equation (2.19) and set $dr/d\tau = 0$, the condition for a circular orbit.

Problem 16 From the geodesic equations (2.14) and (2.17)

$$\frac{dt}{d\tau} = \frac{E}{\left(1 - \frac{2m}{r} \right)} \tag{2}$$

and

$$\frac{d\phi}{d\tau} = \frac{L}{r^2} \tag{3}$$

where τ is the proper time of the orbiting particle. Substituting the expression for L obtained in problem 15 into (3) and integrating over an orbit gives

$$\tau = \frac{2\pi r^{3/2}}{m^{1/2}} \left(1 - \frac{3m}{r}\right)^{\frac{1}{2}}.$$

To obtain the period measured by a distant observer we divide (3) by (2):

$$\frac{d\phi}{dt} = \frac{d\phi/d\tau}{dt/d\tau} = \frac{L(1 - 2m/r)}{r^2 E}.$$

Now substitute for E and L the expressions obtained in problem 15 and integrate over one orbit to get

$$T = \frac{2\pi r^{\frac{3}{2}}}{m^{\frac{1}{2}}}.$$

Problem 17 The energy per unit mass of a particle in a circular orbit is

$$E = \left(1 - \frac{2m}{r}\right)\left(1 - \frac{3m}{r}\right)^{-\frac{1}{2}}.$$

Put $E = 1$, the energy of a particle at rest at infinity, and solve for r to get $r = 4m$. Inserting this value of r into $L = (mr)^{\frac{1}{2}}(1 - 3m/r)^{-\frac{1}{2}}$ gives $L = 4m$ thus $L/E = 4m$. This is an unstable orbit because it lies inside the last stable orbit at $r = 6m$. Note that in Newtonian physics a particle coming in from infinity must loose energy before it can be inserted into a circular orbit.

Problem 18 The velocity v of a particle in a circular orbit at radius r is related to the escape velocity v_{esc} from that radius by

$$v = \frac{v_{\text{esc}}}{\sqrt{2}(1 - 2m/r)^{1/2}}.$$

Clearly when $r = 4m$ then $v = v_{\text{esc}}$. For $r < 4m$ the value of the denominator falls with decreasing r and $v > v_{esc}$.

Problem 19 We evaluate $d\tau/m_0$ in a local inertial frame as we can use special relativity in such a frame. The infinitesimal time-like interval between a pair of events along the world line of a particle having velocity u in flat spacetime is $d\tau = dt(1 - u^2)^{1/2}$. The mass m_0 of the particle is related to its energy and momentum by $m_0 = (E^2 - p^2)^{1/2} = E(1 - u^2)^{1/2}$ as $p = Eu$: note that E here is the energy of the particle not the energy per unit mass. So $d\tau/m_0 = dt/E$. This expression is independent of the velocity of the particle so it also applies to photons. Hence $d\tau/m_0 \rightarrow dt/E_{ph}$ which is finite.

Problem 20 The 4-velocity of a hovering observer at an arbitrary radial coordinate r is $(u^\mu) = \left((1 - 2m/r)^{-\frac{1}{2}}, 0, 0, 0\right)$ and the 4-momentum of a particle at r has covariant components $(p_\mu) = (E, p_1, p_2, p_3)$. To obtain the velocity of the particle with respect to the local hovering observer we evaluate the scalar product $u^\mu p_\mu = \gamma m_0$ in the local frame and in the Schwarzschild frame and equate the results. This gives $m_0 E \left(1 - 2m/r\right)^{-\frac{1}{2}} = m_0(1 - v^2)^{-\frac{1}{2}}$. Solving for v the velocity of the particle gives

$$v^2 = \frac{m_0^2 E^2 - (1 - 2m/r)m_0^2}{m_0^2 E^2}.$$

Now as $m_0 \to 0$, $m_0 E \to E_{ph}$ and the second term in the numerator goes to zero. Thus in the limit $v = 1$, the speed of light. An alternative way to do this is to cancel the m_0 and remember that E is the energy at infinity, so $E^2 = (1 - v_\infty^2)^{-1}$. Now $v^2 = 1 - (1 - v_\infty^2)(1 - 2m/r)$. Set v at infinity equal to the speed of light; then the local speed v is everywhere equal to the speed of light.

Problem 21 This calculation follows closely the calculation in section 2.4.1 We evaluate the scalar product $u^\mu p_\mu = h\nu$, where u is the 4-velocity of the falling spaceship, p the 4-momentum of a photon coming from infinity and $h\nu$ is the energy of the photon in the spaceship frame. Now we evaluate the scalar product in the Schwarzschild frame which is the frame of an observer at infinity and equate it to $h\nu$. The components of the probe's 4-velocity in the Schwarzschild frame are $(u^\mu) = \left(\left(1 - \frac{2m}{r}\right)^{-1}, -\left(\frac{2m}{r}\right)^{1/2}, 0, 0\right)$ and the covariant components of momentum of the inwardly propagating photon are $(p_\mu) = \left(h\nu_\infty, h\nu_\infty \left(1 - \frac{2m}{r}\right)^{-1}, 0, 0\right)$. So finally we get

$$h\nu_\infty \left[1 + \left(\frac{2m}{r}\right)^{1/2}\right]^{-1} = h\nu.$$

The crew can continue to receive signals even when they are inside the horizon. Only as the singularity at $r = 0$ is approached does the signal become infinitely redshifted.

Problem 22 The conditions for a circular orbit are $\frac{dr}{d\lambda} = 0$ and $\frac{d^2r}{d\lambda^2} = 0$. Equation (2.28) gives

$$\left(\frac{dr}{d\lambda}\right)^2 = E_{ph}^2 - \frac{L_{ph}^2}{r^2}\left(1 - \frac{2m}{r}\right).$$

Differentiating we get

$$\frac{d^2r}{d\lambda^2} = \frac{L_{ph}^2}{r^3}\left(1 - \frac{2m}{r}\right) - \frac{2mL_{ph}^2}{r^4}.$$

With $L_{ph}/E_{ph} = 3\sqrt{3}m$ both of these expressions are zero at $r = 3m$. So the light ray is indeed in a circular orbit at this radius.

Problem 23 The apparent diameter of the hole is $3\sqrt{3}(2GM/c^2) = 6.2 \times 10^{10}$ m. So at the Earth it subtends an angle $\sim 51\mu$as.

Problem 24 We start from

$$\left(\frac{dr}{d\phi}\right)^2 = \left(\frac{dt/d\tau}{d\phi/d\tau}\right)^2 = \frac{E^2 - (1 + L^2/r^2)(1 - 2m/r)}{L^2/r^4}.$$

Let $u = 1/r$ then

$$\frac{du}{d\phi} = -\frac{1}{r^2}\frac{dr}{d\phi}.$$

Changing to the new variable in the above expression gives

$$\left(\frac{du}{d\phi}\right)^2 = \frac{E^2}{L^2} - \left(1 + L^2 u^2\right)(1 - 2mu)/L^2,$$

which, on differentiating with respect to ϕ, gives

$$\frac{d^2u}{d\phi^2} + u = \frac{m}{L^2} + 3mu^2.$$

In the Newtonian limit the term $3mu^2$ is zero so it represents the departure from Newtonian physics. Making the substitution $u = l^{-1}(1 + \varepsilon\cos\lambda\phi)$ and collecting together the cosine terms gives

$$\left(-\frac{\varepsilon\lambda^2}{l} + \frac{\varepsilon}{l} - \frac{6m\varepsilon}{l^2}\right)\cos\lambda\phi + \frac{1}{l} = \frac{m}{L^2} + \frac{3m}{l^2}.$$

For a solution to this equation the coefficient of the $\cos\lambda\phi$ term must vanish as there is no ϕ on the right hand side; that is $\lambda = (1 - 6m/l)^{1/2} \sim 1 - 3m/l$ and $1/l = m/L^2 + 3m/l^2 = m/L^2$. In the Newtonian limit $\lambda = 1$ so here $3m/l$ is small for weak fields. In one orbit $\lambda\phi = 2\pi$ so $\phi = 2\pi/(1 - 3m/l) \approx 2\pi + 6\pi m/l$. Therefore in one orbit ϕ has turned through the extra angle $6\pi m/l = 6\pi m^2/L^2$. For a circular or near circular orbit $L^2 = mr/(1 - 3m/r) = mr$ for $r >> m$. So the amount of precession per orbit is $\Delta\phi = 6\pi m/r$. For the planet Mercury, for example, this comes to 43 seconds of arc per century. For a light ray

$$\left(\frac{dr}{d\phi}\right)^2 = \frac{E_{ph} - (L^2/r^2)(1 - 2m/r)}{L^2/r^4}.$$

Setting $u = 1/r$ and differentiating gives

$$\frac{d^2u}{d\phi^2} + u = 3mu^2.$$

In the absence of the term $3mu^2$ the solution to the equation is of the form $u = A\cos(\phi + \alpha)$, which is the equation of a straight line in polar coordinates. Finally setting $u = $ constant in the above equation gives the solution $1/u = r = 3m$.

Problem 25 (a) From the result of problem 12, $\frac{dr}{d\tau} = -(2m/r)^{1/2}$ for a particle falling from rest at infinity. Separating the variables and integrating:

$$\int_0^T d\tau = -\int_{2m}^0 \frac{r^{1/2}}{(2m)^{1/2}} dr$$

from which $T = 4m/3$.

(b) Use $g_{\mu\nu} u^\mu u^\nu = 1$. Inside the horizon this gives

$$1 = -\left(\frac{2m}{r} - 1\right) \dot{t}^2 + \left(\frac{2m}{r} - 1\right)^{-1} \dot{r}^2 - r^2\dot{\theta}^2 - r^2 \sin^2\theta \dot{\phi}^2,$$

where the dot represents differentiation with respect to proper time. For this to be satisfied the positive term on the right hand side must be ≥ 1. The equals sign will give the maximum proper time so

$$\left(\frac{2m}{r} - 1\right)^{-1} \dot{r}^2 = 1.$$

And hence

$$\frac{dr}{d\tau} = \pm \left(\frac{2m}{r} - 1\right)^{\frac{1}{2}}.$$

Taking the minus sign for inward motion and separating the variables gives

$$\int_0^T d\tau = -\int_{2m}^0 (2m/r - 1)^{-\frac{1}{2}} dr.$$

Making the substitution $r = 2m \cos^2\theta$ and integrating gives $T = \pi m$.

(c) Paths of maximum proper time are geodesics.

(d) Use equation (2.19) and set $L = 0$ and $E = 0$ as the conditions for a particle to be at rest at the horizon. This gives

$$\frac{dr}{d\tau} = -\left(\frac{2m}{r} - 1\right)^{\frac{1}{2}}$$

which, as in part (b), integrates to $T = \pi m$. So an astronaut who wishes to survive for as long as possible inside the horizon should fire his rockets outside the horizon to reduce his speed as much as possible before crossing the horizon.

Problem 26 The time spent in pain is $\tau = \frac{2}{3}\left(\frac{\Delta r}{g_E}\right)^{\frac{1}{2}}$. To convert from geometric units to physical units use t(in metres) $= ct$(in seconds) and g(metres^{-1}) $=$

$c^{-2}g(\text{ms}^{-2})$. On making these substitutions the c's cancel so the original expression is valid. Inserting $\Delta r = 2m$ and $g_E = 9.81$ ms^{-2} into the expression above gives $\tau = 0.3$s. The threshold of pain radius is $r_p = \left(\frac{2m\Delta r}{g_E}\right)^{\frac{1}{3}}$ in geometric units. Converting to physical units this becomes $r_p = \left(\frac{2GM\Delta r}{g_E}\right)^{\frac{1}{3}}$, after converting g and setting $m = GM/c^2$. For $M =$ one solar mass $r_p = 3.8 \times 10^3$ km. This is well beyond the horizon of a one solar mass black hole, which is about 3 km. The threshold of pain equals the Schwarzschild radius when $r_p = r_s$. This condition is realised when $M = \frac{c^3}{2G}\left(\frac{\Delta r}{g}\right)^{1/2} = 4.6 \times 10^4$ solar masses.

Problem 27 From the answer to problem 22 the time from crossing the horizon to hitting the singularity for a particle falling from a large distance is $4m/3$. Converting this expression to physical units and inverting to give mass in terms of time T gives

$$M = \frac{3c^3 T}{4G}.$$

For $T = 1$ yr we get $M = 5 \times 10^{12}$ solar masses. Note that only in the last 0.3 seconds will serious deformation occur, but 0.3 s is too short a time in which to suffer any pain.

Problem 28 In Rindler coordinates the metric is $d\tau^2 = \kappa^2\xi^2 dt^2 - d\xi^2$. Transforming to the coordinates (T,X) where

$$T = \xi \sinh \kappa t$$
$$X = \xi \cosh \kappa t$$
$$dT = d\xi \sinh \kappa t + dt\kappa\xi \cosh \kappa t$$
$$dX = d\xi \cosh \kappa t + dt\kappa\xi \sinh \kappa t.$$

Squaring and adding gives

$$dT^2 - dX^2 = \kappa^2\xi^2 dt^2 - d\xi^2.$$

So

$$d\tau^2 = dT^2 - dX^2.$$

Problem 29 On the curve $\xi = 1/a =$ constant

$$T = a^{-1} \sinh \kappa t$$
$$X = a^{-1} \cosh \kappa t.$$

Differentiating T and X with respect to τ gives

$$\frac{dT}{d\tau} = a^{-1}\kappa \cosh \kappa t \frac{dt}{d\tau},$$

and

$$\frac{dX}{d\tau} = a^{-1}\kappa \sinh \kappa t \frac{dt}{d\tau}.$$

Differentiating these expressions again with respect to τ gives the two non zero components of the 4-acceleration:

$$\frac{d^2 T}{d\tau^2} = a^{-1}\kappa^2 \sinh \kappa t \left(\frac{dt}{d\tau}\right)^2 + a^{-1}\kappa \cosh \kappa t \frac{d^2 t}{d\tau^2}$$

$$\frac{d^2 X}{d\tau^2} = a^{-1}\kappa^2 \cosh \kappa t \left(\frac{dt}{d\tau}\right)^2 + a^{-1}\kappa \sinh \kappa t \frac{d^2 t}{d\tau^2}.$$

Now from the Rindler metric with $\xi = 1/a = $ constant

$$\left(\frac{dt}{d\tau}\right)^2 = \frac{1}{\kappa^2 \xi^2} = \frac{a^2}{\kappa^2} \quad \text{and} \quad \frac{d^2 t}{d\tau^2} = 0.$$

So the proper acceleration $A^\mu = (a \sinh \kappa\tau, a \cosh \kappa\tau, 0, 0)$ and $A^\mu A_\mu = -a^2$. So the magnitude of the proper acceleration is a.

Problem 30 We wish to find the limiting value of $\cos \psi$ as $r \to 0$, where

$$\cos \psi = \frac{[1 - (1 - 2m/r) b^2/r^2]^{1/2} - (2m/r)^{1/2}}{1 - (2m/r)^{1/2} [1 - (1 - 2m/r) b^2/r^2]^{1/2}}.$$

The velocity of a particle falling from rest at infinity, $(2m/r)^{1/2}$, has been inserted into equation (2.52). As $r \to 0$ the term $(1 - 2m/r)b^2/r^2$ dominates so

$$\cos \psi \to \frac{[(1 - 2m/r)b^2/r^2]^{1/2}}{(2m/r)^{1/2} [1 - (1 - 2m/r)b^2/r^2]^{1/2}} \to -\left(\frac{r}{2m}\right)^{1/2}.$$

Problem 31 From section 3.2 the event horizon is a surface of constant r and constant t but its equatorial and polar circumferences are not the same. From section 3.2.1 the equatorial circumference of an extreme Kerr black hole is $2\pi(r_+^2 + m^2)/r_+ = 4\pi m$ as $r_+ = m$. From the metric, the polar circumference of this black hole is $\int_0^{2\pi} \rho d\theta = \int_0^{2\pi} \left(r_+^2 + m^2 \cos^2 \theta\right)^{\frac{1}{2}} d\theta = m \int_0^{2\pi} \left(1 + \cos^2 \theta\right)^{\frac{1}{2}} d\theta$. This integral cannot be evaluated in terms of elementary functions. However we can estimate its value as follows. Expanding $(1 + \cos^2 \theta)^{\frac{1}{2}}$ by the binomial theorem gives $1 + \frac{1}{2}\cos^2 \theta - \frac{1}{8}\cos^4 \theta + \frac{1}{16}\cos^6 \theta + ...$ for the integrand. We can evaluate these four integrals: they give respectively $2\pi + \pi/2 - 3\pi/16 + 5\pi/48 = 2.42$. So the circumference is $\sim 2.42\pi m = 7.6m$. This is less than the equatorial circumference.

Problem 32 Putting $m = 0$ into the Boyer Lindquist form of the Kerr metric gives

$$ds^2 = dt^2 - \frac{\rho^2}{r^2 + a^2} dr^2 - \rho^2 d\theta^2 - (r^2 + a^2)\sin^2\theta d\phi^2.$$

We have to show that the coordinate transformation given in the question changes the flat space metric $dx^2 + dy^2 + dz^2$ into the spatial part of the metric given above. Differentiating the x coordinate transformation gives

$$dx = (r^2 + a^2)^{-\frac{1}{2}} r \sin\theta \cos\phi dr + (r^2 + a^2)^{\frac{1}{2}} \cos\theta \cos\phi d\theta - (r^2 + a^2)^{-\frac{1}{2}} \sin\theta \sin\phi d\phi.$$

So

$$\begin{aligned} dx^2 = {}& (r^2 + a^2)^{-1} r^2 \sin^2\theta \cos^2\phi dr^2 + (r^2 + a^2)\cos^2\theta \cos^2\phi d\theta^2 \\ &+ (r^2 + a^2)\sin^2\theta \sin^2\phi d\phi^2 + 2\cos\theta\sin\theta\cos^2\phi dr d\theta \\ &- 2\sin^2\theta\cos\phi\sin\phi dr d\phi - 2(r^2 + a^2)\cos\theta\sin\theta\cos\phi\sin\phi d\theta d\phi. \end{aligned}$$

Proceeding in the same way with the y and z transformations and adding them together gives finally

$$dx^2 + dy^2 + dz^2 = \frac{\rho^2}{(r^2 + a^2)} + \rho^2 d\theta^2 + (r^2 + a^2)\sin^2\theta d\phi^2.$$

Problem 33 We consider the identity $g_{03}^2 - g_{00}g_{03} = \Delta\sin^2\theta$. Substituting for the metric coefficients and extracting the common factor gives

$$\frac{\sin^2\theta}{\rho^4} \left[4m^2 a^2 r^2 \sin^2\theta + \left(\Delta - a^2\sin^2\theta\right) A \right].$$

Inserting the expressions for A and Δ and using $\rho^2 = r^2 + a^2\cos^2\theta$ gives

$$\frac{\sin^2\theta}{\rho^4} \left[4m^2 a^2 r^2 \sin^2\theta + \left(\rho^2 - 2mr\right)\left(\rho^2\left(r^2 + a^2\right) + 2ma^2 r \sin^2\theta\right) \right].$$

Finally the expression inside the square brackets reduces to $\rho^4\Delta$ so the left hand side becomes $\Delta\sin^2\theta$ as required. The other identities follow in a similar manner.

Problem 34 For $r \gg a$

$$g_{00} = \frac{\left(\Delta - a^2\sin^2\theta\right)}{\rho^2} \rightarrow \frac{\Delta}{r^2} \rightarrow \left(1 - \frac{2m}{r}\right)$$

$$g_{03} = \frac{\left(2mar\sin^2\theta\right)}{\rho^2} \rightarrow \frac{\left(2ma\sin^2\theta\right)}{r}$$

$$g_{22} = -\rho^2 \rightarrow -r^2$$

$$g_{33} = -\frac{A\sin^2\theta}{\rho^2} \rightarrow -r^2\sin^2\theta.$$

With these values of the metric coefficients the Kerr metric takes the form in equation (3.11).

Problem 35 The radius of the event horizon is $r_+ = m + (m^2 - a^2)^{1/2}$. If $a > m$ then r_+ becomes complex so there is no event horizon.

Problem 36 At the turning points of a bound orbit $d\theta/d\tau = 0$, so multiplying equation (3.16) through by $\sin^2 \theta_{TP}$ gives

$$Q \sin^2 \theta_{TP} = \left(1 - \sin^2 \theta_{TP}\right) a^2 \left(1 - E^2\right) \sin^2 \theta_{TP} + \left(1 - \sin^2 \theta_{TP}\right) L_z^2.$$

Rearranging gives a quadratic equation in $\sin^2 \theta_{TP}$

$$a^2 \left(1 - E^2\right) \sin^4 \theta_{TP} + \left[Q - a^2 \left(1 - E^2\right) + L_z^2\right] \sin^2 \theta_{TP} - L_z^2 = 0$$

The solution of which is

$$\sin^2 \theta_{TP} = \frac{-\left(Q - a^2 \left(1 - E^2\right) + L_z^2\right) \pm \left[\left(Q - a^2 \left(1 - E^2\right) + L_z^2\right)^2 + 4a^2 \left(1 - E^2\right) L_z^2\right]^{\frac{1}{2}}}{2a^2 \left(1 - E^2\right)}.$$

For a bound orbit ($E < 1$) the term inside the square root is larger than the first term in the numerator, so to ensure that $\sin^2 \theta_{TP}$ is positive we take the plus sign. On taking the square root we see that there are two real values of $\sin \theta_{TP}$, that is two turning points for a bound orbit. Note that $Q = 0$ gives $\sin^2 \theta_{TP} = 1$: this is an orbit in the equatorial plane. The condition $L_z = 0$ gives $\sin^2 \theta_{TP} = 0$, which is an orbit with turning points at $\theta = 0$ and $\theta = \pi$ i.e. an orbit that passes over the two poles.

Problem 37 We have $\omega = \frac{u^3}{u^0} = \frac{g^{30} u_0 + g^{33} u_3}{g^{00} u_0 + g^{03} u_3} = \frac{g^{30}}{g^{00}}$ where the second equality follows because $L = -u_3 = 0$. Also from $u_3 = g_{03} u^0 + g_{33} u^3$ we get

$$\frac{u^3}{u^0} = \omega = -\frac{g_{03}}{g_{33}}.$$

Problem 38 The circumference of the event horizon of a Kerr black hole is $4\pi m$. (See section 3.2.1) Now velocity = circumference/period = $\frac{4\pi m}{2\pi/\omega}$. From equation (3.29)3.29 $\omega = 1/(2m)$ for an extreme black hole. Hence relative to an observer at infinity the velocity = 1 in geometric units, which is the speed of light.

Problem 39 From equation (3.29) the angular velocity of a Kerr black hole is given by $\omega_+ = \frac{a}{2mr_+}$. For $a << m$, $r_+ \sim 2m$ so $\omega_+ \sim a/(4m^2)$. By the usual definition, the moment of inertia I is given by $I = j/\omega_+$. So putting $j = ma$ and substituting for ω_+ gives $I = 4m^3$.

Problem 40 For a particle released from rest at infinity with zero angular momentum $d\theta/d\tau$ is given by equation (3.18). Setting $E = 1$ gives $d\theta/d\tau = 0$. From equation (3.16) $d\phi/d\tau \neq 0$ so the particle moves inwards on a conical spiral of constant half angle.

Problem 41 Setting $a = 0$ in equation (3.37): $\left(\frac{g_{33}}{-\Delta \sin^2 \theta}\right)^{1/2} = \left(1 - \frac{2m}{r}\right)^{-1/2}$ and $\omega = 0$. So the components of (u^μ_{ZAMO}) become $\left((1 - 2m/r)^{-1/2}, 0, 0, 0\right)$ which are the components of a hovering observer in the Schwarzschild metric.

Problem 42 The 4-velocity of a hovering observer is $(u^\mu) = \left(g_{00}^{-1/2}, 0, 0, 0\right)$ so the covariant components are

$$(u_\mu) = \left(g_{00}^{1/2}, 0, 0, g_{03}g_{00}^{-1/2}\right),$$

where $u_3 = g_{03}u^0$ has been used. Now evaluate the scalar product $u^\mu_{ZAMO}u_\mu$ in the Schwarzschild frame and equate it to the energy per unit mass in a local frame. Using equation (3.37) for u^μ_{ZAMO} with $\theta = \pi/2$ we get

$$\left(\frac{g_{33}}{-\Delta}\right)^{\frac{1}{2}} \left(g_{00}^{1/2} + \omega g_{03}g_{00}^{-1/2}\right) = \gamma.$$

Inserting $\omega = -g_{03}/g_{33}$ and using the identity $g_{03}^2 - g_{00}g_{33} = \Delta \sin^2 \theta$ we get

$$\left(\frac{-\Delta}{g_{33}g_{00}}\right)^{\frac{1}{2}} = \gamma = \left(1 - \nu^2\right)^{-\frac{1}{2}}$$

or

$$\nu^2 = 1 + \frac{g_{33}g_{00}}{\Delta}.$$

Substituting for g_{33} and g_{00} and simplifying gives

$$\nu = \frac{2ma}{r\Delta^{1/2}}.$$

Problem 43 For purely radial motion of light the metric becomes

$$0 = g_{00}\left(\frac{dt}{d\lambda}\right)^2 + g_{11}\left(\frac{dr}{d\lambda}\right)^2.$$

But inside the static limit both the metric coefficients are negative so the condition cannot be satisfied.

Problem 44 Let u^μ be the 4-vector of the stationary particle. Then the 3-component of $u_\mu = g_{\mu\nu}u^\nu$ is

$$u_3 = g_{03}u^0 + g_{33}u^3 = g_{03}u^0$$

as $u^3 = d\phi/d\tau = 0$. Now, from the metric, the 4-velocity of a stationary particle is

$$u^\mu = \left(g_{00}^{-1/2}, 0, 0, 0\right).$$

So $u_3 = -L_z = g_{30}g_{00}^{-1/2}$. Finally

$$L_z = -\frac{2mar\sin^2\theta}{\rho\left(\Delta - a^2\sin^2\theta\right)^{1/2}}.$$

The denominator is zero at the static limit surface so $L_z \to \infty$ there. Therefore only a photon can remain static at this location. Inside the static limit surface the angular momentum becomes imaginary so a particle cannot be at rest.

Problem 45 Equation (3.49) written as a quadratic in a is

$$3a^2 \mp 8m^{\frac{1}{2}}r_{ms}^{\frac{1}{2}}a - \left(r_{ms}^2 - 6mr_{ms}\right) = 0$$

so

$$a = \frac{r_{ms}^{\frac{1}{2}}m^{\frac{1}{2}}}{3}\left(\pm 4 \mp \left(\frac{3r_{ms}}{m} - 2\right)^{\frac{1}{2}}\right).$$

To check the signs set $r_{ms} = m$. This gives $a = m$ (the prograde case) with first sign + and second sign $-$. With $r_{ms} = 9m$ we get $a = m$ (the retrograde case) with the two signs reversed. So for a retrograde orbit with $r_{ms} = 7.5m$, $a = 0.48$ as required. For a prograde orbit with $r_{ms} = 4.12m$, $a = 0.53m$.

Problem 46 We are given $r = 4.12m$ for the last stable orbit of a black hole with $a = 0.53m$ and $T = 2\pi\left(\frac{r^{3/2}}{m^{1/2}} \pm a\right)$. Substituting for r and a gives $T = 2\pi m\left[(4.12)^{3/2} + 0.53\right]$. Putting $m = GM/c^2$ to convert to physical units and using $M = 3.7 \times 10^6 M_\odot$ gives $T = 17.1$ minutes.

Problem 47 We use the expression for a as a function of r_{ms} derived in problem 45 to obtain values of r_{ms} for a range of values of a. We then use these values of r in equation (3.48).

Problem 48 Equation (3.44), after substitution for E^2 from equation (3.48), becomes

$$L^2 = \frac{2m}{3r_{ms}}\left(3r_{ms}^2 - a^2\right). \tag{4}$$

Now solving equation (3.49) as a quadratic in a (see problem 45) gives

$$a = \frac{m^{\frac{1}{2}}r_{ms}^{\frac{1}{2}}}{3}\left[\pm 4 + \left(3\frac{r_{ms}}{m} - 2\right)^{\frac{1}{2}}\right].$$

Squaring a and substituting into (4) gives

$$L^2 = \left(\frac{4m^2}{27}\right)\left[12\left(\frac{r_{ms}}{m}\right) - 7 + 4\left(3\frac{r_{ms}}{m} - 2\right)^{\frac{1}{2}}\right].$$

Finally taking the square root gives

$$L = \pm \frac{2m}{3\sqrt{3}} \left[1 + 2 \left(3\frac{r_{ms}}{m} - 2 \right)^{\frac{1}{2}} \right].$$

Problem 49 (a) For the co-rotating case in the extreme Kerr limit $r_{ms} = m$ so $L = 2m/\sqrt{3}$.
(b) For the counter-rotating case $r_{ms} = 9m$ so $L = -22m/(3\sqrt{3})$.

Problem 50 A direct substitution of $a = 0$ and $r = 3m$ gives $b = 0/0$ which is indeterminate. Instead use the expression in the hint:

$$(r/m)^{\frac{3}{2}} - 3\,(r/m)^{\frac{1}{2}} \pm 2\,(a/m) = 0.$$

Solving for a we get $a = \mp(r/m)^{1/2}(r - 3m)/2$. Now substituting this value for a into the expression for b gives $b = (r/m)^{1/2}(r + 3m)/2$. But $r = 3m$ so $b = 3\sqrt{3}m$.

Problem 51 For large r equation (3.53) becomes

$$E^2 = \frac{r\,(r^2 - 2mr)^2}{r^2\,(r^3 - 3mr^2)} = \frac{\left(1 - \frac{2m}{r} \right)^2}{\left(1 - \frac{3m}{r} \right)} \tag{5}$$

which is the Schwarzschild result for the energy of a particle in a circular polar orbit.

Problem 52 A plot of energy per unit mass E against r for a circular polar orbit shows that the minimum value of the specific energy is 0.937 at $r \sim 5.3m$. Therefore the maximum binding energy is $1 - 0.937 = 0.063$ or 6.3 % of the rest mass of a particle.

Problem 53 (a) In geometrical units $\omega_p = 2ma/r^3$. Now $c \times t(\text{seconds}) = t(\text{metres})$ so $\omega(\text{seconds}^{-1}) = c \times \omega(\text{metres}^{-1})$. We also have $m = GM/c^2$ and $a = J/Mc$. So

$$\omega(s^{-1}) = \frac{2GJ}{c^2 r^3}.$$

(b) We have

$$\omega_p = \frac{2ma}{r^3} \times \frac{m^3}{m^3} = \frac{2a}{m^2\,(r/m)^3},$$

so for a given r/m the precessional frequency is inversely proportional to the mass of the black hole squared.

Problem 54 For a polar orbit about the Earth we can use the expression for the precession angular velocity given in SI units in problem 53

$$\frac{d\phi}{dt} = \frac{2GJ}{c^2 r^3},$$

with $J = 0.33 M_\oplus R_\oplus^2 \Omega_\oplus = 0.33 \times 6 \times 10^{24} \times (6.37 \times 10^6)^2 \times 7.3 \times 10^{-5} = 5.86 \times 10^{33}$ Js. So in one year $\Delta\phi = (2 \times 6.7 \times 10^{-11} \times 5.86 \times 10^{33} \times 3.16 \times 10^7)/(9 \times 10^{16} \times (12 \times 10^6)^3) = 1.6 \times 10^{-7}$ radians or 33 milli-arc seconds.

Problem 55 For a polar orbit $L_z = 0$ so equation (3.16) gives for the polar orbit

$$\rho^2 \frac{d\phi}{d\tau} = \frac{2mar E}{\Delta}. \tag{6}$$

For a zero angular momentum particle in the equatorial plane we have $L_z = 0$ and $\theta = \pi/2$ so for this case equation (3.16) gives

$$r^2 \frac{d\phi}{d\tau} = \frac{2mar E}{\Delta}. \tag{7}$$

When $r \gg a$ we have $\rho \to r$ and these two expressions tend to the same value.

Problem 56 The condition $V_r = 0$ gives

$$Q = \left[E^2 \left(r^2 + a^2 \right)^2 - \Delta a^2 E^2 - r^2 \Delta \right] / \Delta.$$

Substituting for E^2 from equation (3.53) gives, after simplification, the required expression. The final part of the question follows immediately on setting $a = 0$. In this limiting case $Q = L^2$.

Problem 57 The Killing vector $(k^\mu) = (1, 0, 0, 0)$ has covariant components $k_\mu = g_{\mu\nu} k^\nu$ so $k_0 = g_{00} k^0 + g_{03} k^3 = g_{00} k^0$ as $k^3 = 0$. So the scalar product $k^\mu k_\mu = g_{00}$. At the static limit surface g_{00} changes from being positive outside to negative inside. Therefore the vector k^μ is spacelike inside the static limit surface.

Problem 58 Setting the index $\mu = 0$ in $p_\mu = g_{\mu\nu} p^\nu$ for a photon gives

$$p_0 = E_{ph} = g_{00} p^0 + g_{03} p^3 = g_{00} \frac{dt}{d\lambda} + g_{03} \frac{d\phi}{d\lambda}$$

or

$$E_{ph} = \frac{dt}{d\lambda} \left(g_{00} + g_{03} \Omega \right), \tag{8}$$

where $\Omega = \frac{d\phi}{dt}$. Now from equation (3.40) $\Omega = \omega \pm \frac{r^2 \Delta^{1/2}}{A}$ for a tangentially projected photon and $\omega = -\frac{g_{03}}{g_{33}}$. Substituting these expressions into (8) and using the identity $g_{03}^2 - g_{00} g_{33} = \Delta \sin^2 \theta$ gives for motion in the equatorial plane

$$E_{ph} = \frac{dt}{d\lambda} \left(\frac{\Delta r^2 \pm 2mar \Delta^{1/2}}{A} \right). \tag{9}$$

The time-like component of $p^\mu = g^{\mu\nu}p_\nu$ is $\frac{dt}{d\lambda}$:

$$\frac{dt}{d\lambda} = \frac{AE_{ph} - 2marL_{ph,z}}{r^2\Delta}.$$

Substituting into (9) and collecting the terms containing E_{ph} on the left hand side gives, after some cancellations,

$$E_{ph} = \frac{2marL_{ph} + |L_{ph}| r^2\Delta^{1/2}}{A}, \tag{10}$$

where L_{ph} is positive for co-rotation and negative for contra-rotation. A particle at the horizon ($\Delta = 0$) has $E_{ph} = \frac{2mar + L_{ph,z}}{(2mar_+)^2} = \frac{aL_{ph,z}}{2mr_+} = \omega_+ L_{ph,z}$. Inserting $\dot{r} = 0$ into equation (3.56) and setting the $r^2\Delta$ term inside the square root to zero, as we are treating a photon, reduces (3.56), after some simplification, to equation (10). If L is negative then equation (10) tells us that inside the ergosphere, where $r^2\Delta^{1/2} < 2mar$, E_{ph} is negative.

Problem 59 The innermost stable orbit for a co-rotating particle around an extreme Kerr black hole is at $r_{ms} = m$. The angular momentum gained through accretion of a mass dm from the last stable orbit of an extreme black hole, from equation (3.64) with $x_{ms} = 1$, is given by $dj = 2mdm$. Integrating gives $j = m^2$.

Problem 60 From equation (3.68) $r_{ms}m = $ constant, so $9m_K^2 = 6m_S^2$ hence $m_K = \sqrt{2/3}m_s$. So the fractional increase in mass of the black hole is $\sqrt{3/2} - 1 = 22.5$ %. To find the quantity of rest mass that has been accreted we employ the method used above. Equation (3.63) with $r_{ms} = 9m_K^2/m$ gives

$$dm_0 = \frac{dm}{\left(1 - \frac{2m^2}{27m_K^2}\right)^{\frac{1}{2}}}.$$

Now integrating

$$\Delta m = a \int_{m_K}^{\sqrt{3/2}m_K} \frac{dm}{(a^2 - m^2)^{1/2}}$$

where $a = \sqrt{27/2}m_K$. Evaluating the integral by using the substitution $m = a\sin\theta$ gives

$$\Delta m_0 = \left(\frac{27}{2}\right)^{\frac{1}{2}} \left[\sin^{-1}\left(\frac{1}{3}\right) - \sin^{-1}\left(\frac{2}{27}\right)^{\frac{1}{2}}\right] m_K$$

so $\Delta m_0 = 0.236m_K$. This is larger than the increase in the mass of the black hole by $0.011m_K$. So 4.7% of the rest mass accreted has been radiated away. Thus, a black hole accreting from a contra-rotating disc will not be as luminous as one accreting from a co-rotating disc.

Problem 61 To answer this question we have to show that the integral of ds is positive around the circle $r = a\delta$, $T = $ constant in the $Z = 0$ plane. Inserting these conditions into the metric in Kerr-Schild coordinates gives

$$ds^2 = -dX^2 - dY^2 - \frac{2ma^3\delta^3}{(a\delta)^4}\left[\frac{-a\delta\left(XdX + YdY\right) - a\left(XdY - YdX\right)}{(a\delta)^2 + a^2}\right]^2. \quad (11)$$

From the coordinate transformations in section 3.18 we get, with $\theta = \pi/2$:

$$dX = \left(-r\sin\tilde{\phi} + a\cos\tilde{\phi}\right)d\tilde{\phi} = -Yd\tilde{\phi} \quad (12)$$

$$dY = \left(r\cos\tilde{\phi} + a\sin\tilde{\phi}\right)d\tilde{\phi} = Xd\tilde{\phi}. \quad (13)$$

Substituting these results into (11) gives

$$ds^2 = \left[-r^2 - a^2 - \frac{2m}{\delta}\left[\frac{-X^2 - Y^2}{(a\delta)^2 + a^2}\right]^2\right]d\tilde{\phi}^2.$$

With delta negative and small ds^2 is positive and so the interval is timelike.

Problem 62 The event horizon is given by $\Delta = r^2 + a^2 + q^2 - 2mr = 0$. Solving this quadratic in r for the radius of the horizon gives $r_+ = m \pm \left(m^2 - a^2 - q^2\right)^{1/2}$. So r_+ is real only when $m^2 > a^2 + q^2$.

Problem 63 (i) We evaluate the scalar product $X^\mu X_\mu$, where $(X^\mu) = (1, 0, 0, \omega)$. Now $X_0 = g_{00}X^0 + g_{03}X^3$ and $X_3 = g_{30}X^0 + g_{33}X^3$ so $X_\mu = (g_{00} + g_{03}\omega, 0, 0, g_{30} + g_{33}\omega)$ and $X^\mu X_\mu = g_{00} + g_{03}\omega + g_{03}\omega + g_{33}\omega^2$. Now $\omega = -g_{03}/g_{33}$ so $X^\mu X_\mu = -(\Delta\sin^2\theta)/g_{33}$. But at the horizon $\Delta = 0$, so X^μ is null at the horizon.

(ii) The metric in ingoing Eddington-Finkelstein coordinates is

$$ds^2 = \left(1 - \frac{2m}{r}\right)dv^2 + 2dvdr - r^2d\tilde{\omega}^2.$$

$X^\mu = (1, 0, 0, 0)$ is null on the horizon so $X^\nu(\nabla_\nu x^\mu) = \kappa X^\mu$ reduces to

$$\nabla_0 X^0 = \kappa.$$

Now $\nabla_\nu X^\mu = X^\mu_{,\nu} + \Gamma^\mu_{\nu\rho}X^\rho$ so $\nabla_0 X^0 = \Gamma^0_{00}X^0 = \kappa$. From the definition of $\Gamma^\mu_{\nu\rho}$ (see equation (1.10)) with $g_{00} = (1 - 2m/r)$ and $g_{01} = 1$ $\Gamma^0_{00} = \frac{1}{2}g^{01}\frac{\partial g_{00}}{\partial r} = \frac{m}{r^2}$. At the horizon, $r = 2m$ so finally $\kappa = 1/(4m)$.

Problem 64 To verify that $m = \kappa A_h/4\pi + 2\omega_+ j$. Substituting for κ, A_h and j :

$$\kappa A_h/4\pi + 2\omega_+ j = \frac{(m^2 - a^2)^{1/2}}{2mr_+} 2m \left[m + (m^2 - a^2)^{1/2} \right] + \frac{2ma^2}{2mr_+}$$

which simplifies to

$$\frac{m(m^2 - a^2)^{1/2}}{r_+} + \frac{m^2}{r_+} = \frac{m \left[m + (m^2 - a^2)^{1/2} \right]}{r_+} = m.$$

This verifies Smarr's formula.

Problem 65 The solution to this problem follows closely the treatment for the increase in surface area due to the capture of a massive particle given in section 4.2.1. Equation (4.7) gives

$$\delta A_h = \frac{16mr_+\pi}{(m^2 - a^2)^{1/2}} \left(\delta m - \omega_+ \delta j \right).$$

For a photon $\delta m = E_{ph}$ and from section 3.10.3 the impact parameter for a photon in a circular orbit around an extreme Kerr black hole is $b = L_{ph}/E_{ph}$ but $b = 2m$ for such a photon so $L_{ph}/E_{ph} = 2m$ and the change in angular momentum $\delta j = 2m E_{ph}$. Using these results the change in area becomes

$$\delta A_h = \frac{16m\pi E_{ph}}{(m^2 - a^2)^{1/2}} \left(r_+ - a \right) = \frac{16m\pi E_{ph}}{(m^2 - a^2)^{1/2}} \left[(m^2 - a^2)^{1/2} + \varepsilon \right]$$

as $r_+ - a = (m^2 - a^2)^{1/2} + \varepsilon$. (See section 4.2.1.) So finally the change in area as $a \to m$ is

$$\delta A_h = 16m\pi E_{ph}.$$

So the area increases.

Problem 66 Starting from equation (4.5) $A_h = 8\pi m \left[m + (m^2 - a^2)^{1/2} \right]$, rearrange and square to get $\left(\frac{A_h}{8\pi m} - m \right)^2 = m^2 - a^2$. Evaluating the left hand side and putting $a = j/m$ gives $\left(\frac{A_h}{8\pi m} \right)^2 + \frac{j^2}{m^2} = \frac{2A_h}{8\pi}$. Finally multiply though by $4\pi m^2/A_h$ to get $\frac{A_h}{16\pi} + \frac{4\pi j^2}{A_h} = m^2$.

Problem 67 From section 4.2.4 $m_{rot} = j^2 \left(A_h/4\pi \right)^{-3/2}$, $j = I\omega_+$ and for $a \ll m$, $A_h \sim 16\pi m^2$ so $m_{rot} = j^2 \left(A_h/4\pi \right)^{-3/2} = I^2\omega^2/8m^3 = (I/8m^3)I\omega_+^2 = \frac{1}{2} I\omega_+^2$.

Problem 68 The total area of the event horizons of the two Schwarzschild black holes before the merger is

$$A_{initial} = 2 \times 4\pi(2m_i)^2 = 32\pi m_i^2.$$

The area of the resultant black hole is

$$A_{final} = 4\pi(2m_f)^2 = 16\pi m_f^2.$$

To maximise the energy dissipated in the merger the area of the final black hole must be as low as the Hawking area theorem allows, which is $A_{final} = A_{initial}$. So equating areas gives

$$M_f = \sqrt{2}m_i.$$

Thus the fraction of the initial mass lost in the merger is $1 - 2^{-1/2} = 0.29$.

Problem 69 The total horizon area before the merger is

$$A_i = 2 \times 4\pi(m_K)^2$$

and the area of the final Schwarzschild black hole is

$$A_f = 4\pi(2m_s)^2.$$

Equating the area before to the area after gives $m_K = m_S$. So half of the initial mass has been radiated away.

Problem 70 In the limit as $r \to \infty$ the first term inside the curly brackets reduces to ω^2 and the other two terms tend to zero. So the Teukolsky equation simplifies to

$$\frac{d^2u}{dr_*^2} = -\omega^2 u,$$

which has equation (4.24) as its solution. In the limit as $r \to r_+$, $\Delta \to 0$ so in this case the Teukolsky equation reduces to

$$\frac{d^2u}{dr_*^2} = \frac{\left[(r_+^2 + a^2)\,\omega - aM\right]^2 u}{(r_+^2 + a^2)^2} = -(\omega - \omega_+ M)^2 u$$

as $(r_+^2 + a^2) = 2mr_+$ and $\omega_+ = a/2mr_+$. Thus in this limit the Teukolsky equation has equation (4.25) as its solution. Putting $a = 0$ in the Teukolsky equation gives

$$\frac{d^2u}{dr_*^2} + \omega^2 u - \frac{l(l+1)(1 - 2m/r)u}{r^2} - \frac{(1 - 2m/r)^2}{r^2} - \frac{d}{dr_*}\left(\frac{(1 - 2m/r)}{r}\right)u = 0.$$

As

$$\frac{d}{dr_*} = \frac{\Delta}{r^2 + a^2}\frac{d}{dr}$$

the last term becomes $(1 - 2m/r)^2/r^2 - 2m(1 - 2m/r)/r^3$; so finally the Teukolsky equation reduces to

$$\frac{d^2u}{dr_*^2} + \omega^2 u - \frac{(1 - 2m/r)[l(l+1) + 2m/r]}{r^2} = 0$$

which is the Regge-Wheeler equation.

Problem 71 We are given that $W = u'v - v'u = $ constant for any two independent solutions u and v of a second order differential equation. A solution of equation (4.17) in the limit as $r_* \to -\infty$ is $u \sim e^{-i\omega r_*}$ (equation 4.18). Another solution is $v = u^*$. So differentiating u and v with respect to r_* gives $u' = -i\omega e^{-i\omega r_*}$ and $v' = i\omega e^{i\omega r_*}$. Hence $W = -2i\omega$. In the limit as $r_* \to +\infty$ the solutions to (4.17) takes the form $u \sim a_{out}e^{i\omega r_*} + a_{in}e^{-i\omega r_*}$ and $v = u* \sim a_{out}e^{-i\omega r_*} + a_{in}e^{i\omega r_*}$. Differentiating with respect to r_* to get u' and v' we get a second expression for W:

$$W = 2i\omega \left(a_{out}^2 - a_{in}^2 \right).$$

Equating these two expressions gives $1 + a_{out}^2 = a_{in}^2$ the relation quoted in the text as equation (4.20).
In a similar way we can repeat this calculation for the limiting solutions of the Teukolsky equation to get the relation (4.26).

Problem 72 For a black hole radiating as a black body at temperature T we can write $kT \sim h\nu = hc/\lambda$ where λ is the wavelength at the peak of the blackbody spectrum. Now λ will be of order oft he dimensions of the black hole (the only length scale in the problem), hence of order the Schwarzschild radius. So substituting $\lambda = 2GM/c^2$ into the above expression gives $T \sim hc^3/2kGM$.

Problem 73 To show that $U_{\omega'}^{(+)} = \frac{1}{2(\pi|\omega'|)^{1/2}} e^{-i|\omega'|\eta} \xi^{i\omega'/a}$ is a solution of the equation

$$-\frac{1}{a^2\xi^2} \frac{\partial^2 \Phi}{\partial \eta^2} + \frac{1}{\xi} \frac{\partial \Phi}{\partial \xi} + \frac{\partial^2 \Phi}{\partial \xi^2} = 0.$$

The factor $2(\pi |\omega'|)^{-1/2}$ will cancel out so we can ignore it. Differentiating $U_\omega^+ = e^{-i|\omega'|\eta} x i^{i\omega'/a}$ twice with respect to η gives

$$\frac{\partial^2 U}{\partial \eta^2} = i^2 \omega'^2 e^{i\omega'\eta} \xi^{i\omega'/a}.$$

Differentiating $U_\omega^{(+)}$ with respect to ξ gives

$$\frac{\partial U}{\partial \xi} = e^{-i\omega'\eta} \left(i\omega'/a \right) \xi^{-1+i\omega'/a},$$

and differentiating a second time gives

$$\frac{\partial^2 U}{\partial \xi^2} = e^{-i\omega'\eta} \left(i\omega'/a \right) \left(i\omega'/a - 1 \right) \xi^{-2+i\omega'/a}.$$

Now substituting these three partial derivatives into the wave equation the quantity $e^{-i|\omega'|\eta}\xi^{-2+i\omega'/a}$ cancels leaving $-\frac{i^2\omega'^2}{a^2} + \frac{i\omega'}{a} + \frac{i^2\omega'^2}{a^2} - \frac{i\omega'}{a} = 0$ so equation (4.34) is a solution of equation (4.33).

Problem 74 From the definition of r_* given in equation (4.16)

$$\frac{r_*}{2m} - 1 = \frac{r}{2m} + \log\left(\frac{r}{2m} - 1\right) - 1.$$

So near $r = 2m$

$$\frac{r_*}{2m} - 1 \sim \log\left(\frac{r}{2m} - 1\right)$$

and

$$\frac{r}{2m} - 1 \sim 1 - \frac{2m}{r} \sim \kappa^2 \xi^2$$

where the second equality comes from equation (2.39). Combining (74) and (74) gives

$$\kappa^2 \xi^2 \sim \exp\left(\frac{r_*}{2m} - 1\right).$$

Now,

$$U = -\exp\left(\frac{-u}{2m}\right)$$

and $u = t - r_*$. So

$$U = -e^{r_*/4m} e^{-\kappa t}.$$

Using (74) and $\kappa = 1/4m$ we get $U \propto \xi e^{-\kappa t}$.
Similarly the Kruskal coordinate $V = \exp(v/4m)$ near to $r = 2m$ becomes $V \propto \xi e^{\kappa t}$.
The transformation from Rindler coordinates (t, ξ) to Minkowski null coordinates
$u = T - X$ and $v = T + X$ is given by $u = \xi e^{-\kappa T}$ and $v = \xi e^{\kappa T}$ using (2.41). Thus
near $r = 2m$ the metric in Kruskal coordinates has approximately the Minkowski
form in null coordinates.

Problem 75 The entropy of a black hole is given by $S = \frac{kc^3 A_h}{4\hbar G}$, where A is the area
of the horizon. For a spherical black hole the horizon area is $4\pi(2m_{pl})^2$ in geometrical
units. In physical units this becomes $4\pi(2GM_{pl}/c^2)^2$ and $M_{pl} = (\hbar c/G)^{1/2}$. So $A_h = 16\pi\hbar G/c^3$ and hence $S = 4\pi k$.

Problem 76 Substituting $M = 2 \times 10^{30}$ kg into the equation given in the text gives
$t_h = [8.33 \times 10^{19} \times (2 \times 10^{30}/10^{12})^3]/3.16 \times 10^7 = 2.2 \times 10^{67}$ years.

Problem 77 A black hole with a mass of 1.6×10^{16} kg has a lifetime of $t_h \sim 2.5 \times 10^9 \times (1.6)^3$ y or 1.02×10^{10} y. Assume a typical gamma ray energy is 1 MeV. This
corresponds to a temperature of $\sim 10^{10}$ K. Using $T_h \sim 10^{23}/M$ K gives $M \sim 10^{13}$ K
so such a black hole will have an age of order the age of the universe.

Problem 78 We have, $kT = \frac{\hbar c^3}{8\pi GM}$. Setting $M = (G/\hbar c)^{-\frac{1}{2}}$, the Planck mass, gives
$kT = \frac{1}{8\pi}\left(\frac{\hbar c^5}{G}\right)^{\frac{1}{2}}$ so $T = 7.78 \times 10^7$ K. We also have $S = \frac{kc^3}{4\hbar G}A_h$ and $A_h = 1m^2$ so
$S = k\left(\frac{c^3}{4\hbar G}\right) = 1.3 \times 10^{46}$ JK^{-1}.

Problem 79 The relationship between M and T is $M = \frac{\hbar c^3}{8\pi GkT}$ (see section 4.6.1) so

$$\frac{d(Mc^2)}{dT} = -\frac{\hbar c^5}{8\pi GkT^2}.$$

Expressing T in terms of M the result follows.

Problem 80 Let $x = M_{eq}/M_i$ and find the value of x that maximises $(1-x)x^4$ by differentiating with respect to x and setting to zero. This gives

$$-x^4 + 4(1-x)x^3 = 0.$$

Thus $x = 4/5$. Substituting this value of x back in to the original expression gives $(1 - 4/5)(4/5)^4 = 0.82$.

Problem 81 To obtain the surface gravity of a Reissner-Nordstrom black hole we find the acceleration of a hovering observer by repeating the steps in section 2.8.1 for the Schwarzschild case, only this time with $g_{00} = g_{11}^{-1} = (1 - 2m/r + q^2/r^2)$. This gives for the acceleration $a = (m/r^2 - q^2/r^3)(1 - 2m/r + q/r^2)^{-\frac{1}{2}}$. To get the acceleration measured by an observer at infinity we use equation (2.36) from section 2.8.2. with the Reissner-Nordstrom red shift factor replacing the Schwarzschild factor. This gives for the surface gravity $\kappa = \frac{(mr_+ - q^2)}{r_+^3}$, where $r_+ = m + (m^2 - q^2)^{\frac{1}{2}}$. An extreme Reissner-Nordstrom black hole has $q = m$ and $r_+ = m$ hence $\kappa = 0$. And so $T = \hbar\kappa/(2\pi ck) = 0$.

Problem 82 The metric is

$$d\tau^2 = dt^2 - dl^2 - (b_0^2 + l^2)(d\theta^2 + \sin^2\theta d\phi^2).$$

In the $\theta = \pi/2$ plane at constant t the spatial metric is

$$dL^2 = dl^2 + (b_0^2 + l^2)d\phi^2. \tag{14}$$

The metric of Euclidean space is

$$dL^2 = dz^2 + dr^2 + r^2 d\phi^2.$$

On the surface $z = z(r)$ this becomes

$$dL^2 = \left[1 + \left(\frac{dz}{dr}\right)^2\right]dr^2 + r^2 d\phi^2. \tag{15}$$

Equations (14) and (15) are the same if

$$\left(1 + z'^2\right)dr^2 = dl^2 \quad \text{and} \quad r^2 = b_0^2 + l^2.$$

Thus

$$1 + z'^2 = \left(\frac{dr}{dl}\right)^2 = \left(\frac{r}{l}\right)^2.$$

Integrating this equation gives

$$z = b_0 \cosh^{-1}\left(\frac{r}{b_0}\right).$$

Problem 83 Let the falling body have 4-velocity components u^μ. Since the metric is static we have, as usual, $u_0 = E$. Now, a hovering (stationary) observer has 4-velocity $u_H^0 = dt/d\tau = e^{-\Phi}$ so

$$u_\mu u_H^\mu = E e^{-\Phi} = \gamma = (1 - v^2)^{-1/2}$$

from which

$$v = \left(1 - \frac{e^{2\Phi}}{E^2}\right)^{\frac{1}{2}}.$$

The speed v is that measured by the local stationary observer, hence $v =$ (proper distance)/(proper time of the local observer) $= dl/e^\Phi dt$. The time to traverse the wormhole is obtained by integration.

Problem 84 Starting from the metric

$$d\tau^2 = dt^2 - \frac{dr^2}{1 - b_0^2/r^2} - (b_0^2 + l^2)(d\theta^2 + \sin^2\theta d\phi^2)$$

we set $\theta = \pi/2$ and $d\tau^2 = 0$. The constants of the motion are $u_0 = E_{ph}$ and $u_3 = -L_{ph}$. Then $g_{\mu\nu} u^\mu u^\nu = 0$ gives

$$\left(\frac{dr}{d\lambda}\right)^2 = E_{ph}^2 \left[1 - \frac{L_{ph}^2}{E_{ph}^2 r^2}\right]\left(1 - \frac{b_0^2}{r^2}\right).$$

For $L_{ph}/E_{ph} > b_0$ the first bracket on the right is zero for $r > b_0$. This is the point of closest approach: the radial coordinate of the light ray increases past this turning point and the light ray escapes to infinity. For $L_{ph}/E_{ph} < b_0$ the light ray enters the wormhole. The dividing value gives the impact parameter $L/E = b_0$. The capture cross section is then πb_0^2 because r is an area coordinate.

Problem 85 The 4-velocity of a static observer is $u^\mu = (e^{-\Phi}, 0, 0, 0)$ and the 4-acceleration is

$$a^\mu = \frac{Du^\mu}{d\tau} = \frac{du^\mu}{d\tau} + \Gamma^\mu_{\rho\sigma} u^\rho u^\sigma.$$

We have $a^\mu u_\mu = a^0 u_0 = 0$ so $a^0 = 0$. Also, $a^2 = a^3 = 0$ by symmetry. Thus

$$a^1 = \Gamma^1_{00}(u^0)^2$$
$$= \frac{1}{2}g^{11}\left(-\frac{\partial e^{2\Phi}}{\partial r}\right)(u^0)^2$$
$$= -\left(1 - \frac{r}{r}\right)\Phi'.$$

The acceleration a is now obtained from $a^2 = a^\mu a_\mu$ as

$$a = \left(1 - \frac{b}{r}\right)^{\frac{1}{2}}\Phi'.$$

Problem 86 We have $u^\mu v_\mu = \gamma$ and $g^{\mu\nu}v_\mu v_\nu = 1$ from which the result follows with some simple algebra.

Problem 87 If $\rho > 0$ and $p = -\rho$ them $p + \rho = 0$ which is consistent with the weak energy condition. In the frame of an observer moving at 4-velocity v^μ relative to the fluid the energy density is

$$T^{0'0'} = \gamma^2(\rho + v^2 p) = \gamma^2(1 - v^2)\rho = \rho = T^{00}. \tag{16}$$

Problem 88 As both stars are in circular orbits, from Newton's second law,

$$\frac{v_s^2}{r_s} = \frac{GM_d}{r^2}. \tag{17}$$

From the definition of the centre of mass of the system, $M_s r_s = M_d r_d$, therefore the distance r between the two stars is

$$r = r_s\left(1 + \frac{M_s}{M_d}\right). \tag{18}$$

The orbital period of the system is

$$P = 2\pi r_s/v_s. \tag{19}$$

Combining (17) and (18) gives $r_s = \frac{GM_d}{v_s^2(1+M_s/M_d)^2}$ and substituting for r_s from (19) gives $\frac{Pv_s^3}{2\pi G} = \frac{M_d}{(1+(M_s/M_d)^2)}$. But v_s is the velocity in the plane of the orbit; this is related to the line of sight velocity by $V_s = v_s \sin\phi$, so finally

$$\frac{PV_s^3}{2\pi G} = \frac{M_d \sin^3\phi}{(1 + M_s/M_d)^2}.$$

Problem 89 From problem 15 $t = 2\pi(r^3/m)^{\frac{1}{2}}$. Setting $r = 6m$, the radius of the last stable orbit, $t = 2 \times 6^{3/2}\pi m$. Converting to physical units gives $t = 2 \times 6^{3/2}\pi GM/c^3$ $= 28.3$ minutes.

References

Chapter 1

Hartle J 2003 *Gravity, An Introduction to Einstein's General Relativity* (Addison Wesley)

Kenyon I R 1990 *General Relativity* (Oxford University Press)

Will C 1993 *Theory and Experiment in Gravitational Physics* (Cambridge University Press, Cambridge UK)

Chapter 2

Doeleman S S et al 2008 Nature **455** 78

Frolov V P and Novikov I D 1998 *Black Hole Physics* (Kluwer, Dordrecht)

Oppenheimer J R and Snyder H 1939 Phys Rev **56** 45

Taylor J H 1994 Rev Mod Phys **66** 711

Taylor E F and Wheeler J A 2000 *Exploring Black Holes* (Addison Wesley Longman)

Thorne K S, Price R H and Macdonald D A 1986 *Black Holes: The Membrane Paradigm* (Yale University Press, New Haven)

Chapter 3

Bardeen J M et al 1972 Astrophys J **178** 347

Bardeen J M 1970 Nature **226** 64

Ciufolini I et al 1998 Science **279** 2100

Frolov V P and Novikov I D 1998 *Black Hole Physics* (Kluwer, Dordrecht/London)

Lynden-Bell D 1978 Physica Scripta **17** 185

Begelman M and Rees M 1998 *Gravity's Fatal Attraction* (Scientific American Library)

Thorne K S 1974 Astrophys J **191** 507

Townsend P *Black Holes,* www.damtp.cam.ac.uk/lecturenotes/BH.ps (unpublished lecture notes)

Chapter 4

Bekenstein J D 1975 Phys Rev **D12** 3077

Bekenstein J D 1980 *Black Hole Thermodynamics* Physics Today **24**

Birrell N D and Davies P C W 1982 *Quantum fields in Curved Spacetime* (Cambridge University Press)

Chandrasekhar S 1983 *The Mathematical Theory of Black Holes* (Oxford University Press)

Custodio and Horvath 2003 Am J Phys **71** 1237

Frolov V P and Novikov I D 1998 *Black Hole Physics* (Kluwer, Dordrecht/London)

Futterman J A H, Handler F A and Matzner R A, 1988 *Scattering from Black Holes* (Cambridge University Press, Cambridge UK)

Hawking S W 1974 Nature **248** 30

Hawking S W 1976 Phys Rev **D13** 191

Keifer C 1999 in *Classical and Quantum Black Holes* eds P Fré, V Gorini, G Magli and U Moschella (Institute of Physics, Bristol)

Takagi S 1986 *Vacuum Noise and Stress Induced by Uniform Acceleration*, Prog Theor Phys Supp **88** 1

Townsend P K *Black Holes,* www.damtp.cam.ac.uk/lecturenotes/BH.ps (unpublished lecture notes)

Unruh 1976 Phys Rev **D14** 870

Chapter5

Bordag M, Mohideen U and Mostepanenko V M 2001 Physics Reports **353**

Jensen B, McLaughlin J and Ottewill A 1988 Class. Quantum Grav. **5** L187-L189

Morris M S and Thorne K S 1988 Am J Phys **56** 395

Morris M S, Thorne K S and Yurtsever U 1988 Phys Rev Lett **61** 13

Thorne K S 1994 *B*lack Holes and Time Warps: Einstein's Outrageous Legacy (Picador)

Visser M 1995 *Lorentzian Wormholes* (AIP)

Chapter 6

Bardeen J M and Pettersen J A 1975 Astrophys J **105** L65

Barkana L H and Loeb A 2003 Nature **421** 341

Burgay M et al 2003 Nature **426** 531

Cottam J et al 2002 Nature **420** 51

Gebhardt K, Rich R M and Ho L 2002 Astrophys J **L41** 578

Genzel R et al 2003 Nature **425** 934

Ghez A et al 2008 Astrophys J **689** 1044

Begelman M and Rees M 1998 *Gravity's Fatal Attraction* (Scientific American Library)

Hartle J 2003 *Gravity, An Introduction to Einstein's General Relativity* (Addison Wesley)

Hawking S W and Ellis G F R 1973 *The Scale Structure of Spacetime* (Cambridge University Press)

Hjorth J et al 2003 Nature **423** 847

MacGibbon J and Carr B 1991 Astrophys J **371** 447

McLintock J E et al 2003 Astrophys J **593** 435

Melia F and Falcke H 2001 Ann Rev Astron Astrophys **39** 342

Narayan R, Garcia M and McClintock J E 2002 in Proc IX Marcel Grossmann Meeting eds V Gurzadyan, R T Jantzen and R Ruffini (World Scientific) p405

Orosz J A et al 2007 Nature **449** 872

Price R H 1972 Phys Rev **D5** 2410

Taylor J H 1994 Rev Mod Phys **66** 711

Van Der Marel R P et al 2002 Astron J **124** 3255

Zhang S N and Cui W 1998 Astrophys J **482** L155

Bibliography

The following is a selection of books on black hole physics at various levels.

Thorne K S 1994 *Black holes and time warps* (Papermac): gives a non-mathematical historical survey and includes the problems of time travel.

Luminet J-P 1987 *Black holes* (Cambridge University Press): a semi-popular exposition.

Taylor E F and Wheeler J A 2000 *Exploring black holes* (Addison Wesley Longman): an introduction for those with a basic knowledge of general relativity but without using tensor calculus.

Lightman A P et al 1974 *Problem book in relativity and gravitation* (Princeton): problems for advanced students (with outline answers).

Fre P et al ed 1999 *Classical and quantum black holes* (Institute of Physics): monograph on current developments in black holes and superstrings with an introductory first chapter.

Chandrasekhar S 1983 *The Mathematical theory of black holes* (Oxford): comprehensive mathematical treament of classical black hole spacetimes.

Frolov V P and Novikov I D 1998 *Black hole physics* (Dordrecht; London: Kluwer): detailed survey of black including quantum and astrophysical aspects.

Frank J, King A R and Raine D J 2002 *Accretion power in astrophysics* (Cambridge University Press): astrophysics of gas flows round compact objects.

Index